MEETING-THE-CHALLENGE SERIES

The Ohio Graduation Test

THE SKILLS YOU NEED
MATHEMATICS

ORANGE FRAZER PRESS • P.O. BOX 214 • WILMINGTON OH 45177 • 800.852.9332

ISBN 1-882203-88-7
Copyright 2003 by Shari Wolf

No part of this publication may be reproduced in any material form (including photocopying or storing in any medium by electronic means and whether or not transiently or incidentally to some other use of this publication) without the written permission of the copyright holder except in accordance with the provisions of the Copyright, Designs and Patents Act 1988.

Additional copies of *Meeting the Challenge: The Ohio Graduation Test—Mathematics Study Guide*, or any other Orange Frazer Press books may be ordered directly from:

Orange Frazer Press, Inc.
Box 214
37 ½ West Main Street
Wilmington, Ohio 45177

Telephone 1.800.852.9332 for price and shipping information
Web Site: www.orangefrazer.com

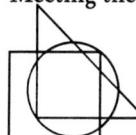 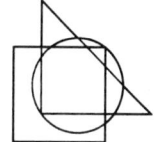

About the Author

Shari Wolf became involved in writing mathematics study materials and in leading tutoring sessions when Ohio's 9th Grade Proficiency Test was implemented in the early 1990's. She also organized formal classes in a large regional high school to help students successfully complete this last step toward education. Since 1995, Shari has continued her involvement by authoring all of the math proficiency materials published by Hollandays Publishing Corporation.

Shari designed Meeting the Challenge ... Mathematics Study Guide to provide the classroom teacher with a plan and materials that will prepare students for the Ohio Graduation Test. This test, based on 10th grade math skills will require students to take their mathematics knowledge to a whole new level of competence.

Shari Wolf earned her Bachelor of Science degree in Mathematics Education from Wright State University and her Masters of Science degree in Guidance from the University of Dayton. After spending fourteen years teaching high school students on all levels of the traditional secondary mathematics curriculum, she then accepted the challenge of developing an applied mathematics curriculum in the Ohio vocational schools setting. Shari's focus throughout her teaching career has been to develop curriculum and materials to help students succeed in mathematics.

You can communicate directly with Shari through her website at www.gradtestmathhelp.com

Acknowledgments

I must thank so many special people who have supported and encouraged me throughout this experience. I believe this project has been a work in progress throughout my entire life as an educator. Thank you Mrs. Aloysia Mumma, my 10th grade geometry instructor at Patterson Cooperative High School in Dayton, Ohio, and Dr. Carl Benner, my education advisor at Wright State University, who together nurtured a young and enthusiastic math teacher and taught her not to back away from any challenge. Thank you to my fellow mathematics instructors from Trotwood Madison High School and the Miami Valley Career Technology Center for their professionalism and support throughout my career. Thank you to Ramona Cann, Academics Supervisor at MVCTC, who placed such confidence in my curriculum development skills that allowed me the opportunity to lead mathematics instruction in an exciting new direction.

I must thank my brother, Steve, for his positive nuts and bolts approach to this opportunity, and to my extended music family for their understanding of the importance of this project to me. Thank you Sharyn McCrumb for letting me know that I, too, could be a writer. Thank you Beverly Smith for giving me the opportunity to continue helping students and teachers and to keep mathematics education in my life through the creation of proficiency and now, the Ohio Graduation Test materials.

You would not be reading this page right now if it were not for my wonderful and capable friends, Joyce Harrison and Linda Hellinger. Linda has been my mathematics and verbiage safety net for the last eighteen months. She retired as an elementary math educator just in time to join us on this project. Her proofreading, math editing and rewriting skills have made her an invaluable member of our team. THANK YOU to Joyce Harrison for her excellent design and formatting skills that have really set the tone for learning on every page of this study guide in the most clear and concise manner possible.

Finally, thank you to Orange Frazer Press for supporting and publishing this study guide. All of us hope that our work and this study guide will be that critical tool to help students pass the mathematics portion of the Ohio Graduation Test and to help educators prepare for it.

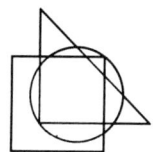 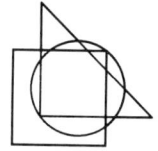

Meeting the Challenge ... Mathematics Study Guide

TABLE OF CONTENTS

Preparing for the OGT Mathematics Test vii

UNIT 1: Number, Number Sense and Operations
- Number and Number Systems
  - Lesson #1   EQUIVALENT NUMBERS 1
- Computation, Estimation and Meaning of Operations
  - Lesson #2   COMPUTING AND ESTIMATING 9
  - Lesson #3   RATIO AND PROPORTION 19
- Mathematical Processes – Communicating with Math 29

UNIT 2: Measurement
- Measurement Units
  - Lesson #4   MEASUREMENT 31
- Use Measurement Techniques and Tools
  - Lesson #5   LENGTH, PERIMETER, AREA, SURFACE AREA and VOLUME 37
- Mathematical Processes – Visualizing to Solve Problems 53

UNIT 3: Geometry and Spatial Sense
- Characteristics and Properties, Transformations and Symmetry
  - Lesson #6   ANGLES AND LINES 55
  - Lesson #7   CONGRUENT AND SIMILAR FIGURES 67
- Spatial Relationships, Visualization and Geometric Models
  - Lesson #8   GEOMETRIC FIGURES AND OBJECTS 79
- Mathematical Processes – Using Diagrams to Solve Problems 91

UNIT 4: Patterns, Functions and Algebra
- Use Patterns, Relations and Functions
  - Lesson #9   REPRESENTING MATHEMATICAL RELATIONSHIPS 93
- Use Algebraic Representations and Analyze Change
  - Lesson #10   ALGEBRAIC EXPRESSIONS AND FORMULAS 103
  - Lesson #11   LINEAR EQUATIONS, INEQUALITIES AND SYSTEMS 117
  - Lesson #12   CREATING AND ANALYZING GRAPHS 129
- Mathematical Processes – Finding Answers in Different Ways 143

UNIT 5: Data Analysis and Probability
- Data Collection and Statistical Methods
  - Lesson #13   TABLES, CHARTS AND GRAPHS 145
  - Lesson #14   MEASURES OF CENTRAL TENDENCY 157
- Probability
  - Lesson #15   PROBABILITY 165
- Mathematical Processes – Using Mathematics 173

Mathematics Formula and Fact Sheet 175
Measurement Study Sheet 176
Glossary 177

> This Study Guide was prepared in accordance with the Ohio Academic Content Standards
> and Benchmarks to assist students in preparation for the Ohio Graduation Test.

© 2003 Orange Frazer Press. All Rights Reserved.

© 2002 Orange Frazer Press. All Rights Reserved.

PREPARING FOR THE OGT MATHEMATICS TEST

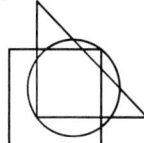
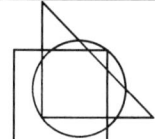

The six major sections of Ohio's Academic Content Standards for Mathematics that will be tested on the OGT Mathematics Test are:

- Number, Number Sense and Operations
- Measurement
- Geometry and Spatial Sense
- Patterns, Functions and Algebra
- Data Analysis and Probability
- Mathematical Processes

There will be 50 questions on the test. Questions from all these areas will be included on the test. The questions are written like those you answered on the 4th and 6th grade proficiency tests. There are multiple-choice questions (worth 1 point each), short answer questions (worth 2 points each), and extended response (essay-type) questions (worth 4 points each). Most of the questions on the test are multiple-choice, but remember, one short answer question is worth *two* multiple-choice questions. Likewise, one extended response question is worth *four* multiple-choice questions.

Study Strategy

You must study for all parts of the test since the points are fairly equally distributed among the six sections. If you like arithmetic and study only arithmetic, you will not be ready for the measurement, algebra, geometry, and data analysis questions. Don't give up if one section is difficult for you. For example, you could do poorly on geometry and still pass the test by having high scores in the other areas.

Take this test seriously. Beginning six weeks before the test, study ½ hour per day using this study guide, the flash cards, and/or the software. This should be in addition to the time you spend in class. Do yourself a favor and do your best the *first* time you take the test. You must pass it to graduate, so get prepared and pass it as soon as you can.

Test Day Strategy

Stay at home the night before the test and go to bed early. Get a good night's sleep. If you have a job, let your boss know several weeks in advance that you need to be off work the night before your graduation test. Many students fail a part of the test they could easily pass if they were wide-awake.

Eat breakfast. Just like your car, your body needs fuel to get going. Take a minute to eat something.

Use your best handwriting or printing. The people who grade these tests read thousands of them. Illegible writing could influence the outcome of your score. On both the short answer and the extended response questions, make sure that your explanation is clear. Show your work completely and write sentences that cover the key points using mathematics vocabulary. Make sure you mark your multiple-choice selection clearly and erase completely. Make it easy for your grader to read your paper.

Verify your calculations by working out the problems manually (showing all your work) and by using your calculator. Then make sure that your answers make sense and are realistic. I cannot emphasize these ideas enough.

Your Reading Skills and the Graduation Test

Many questions in all sections of the graduation test are nothing more than reading questions. For instance, on the mathematics test you will be asked to read and analyze a variety of word problems. You really don't need to know much, if anything, about the application ahead of time to answer the question. All you must do is *read* the information given to you and use your mathematics skills to answer the question.

Here is some specific advice for reading the graduation test:

1. Read each problem and the answers completely before starting. Many students have said they did not read the word problems on past graduation tests because the reading was boring and they hated word problems. To avoid reading they just guessed at the answers. Boring or not, read all the problems and read them carefully. Draw the geometric figures. Make sketches. Remember, your diploma is at stake.

   Once you have solved a problem, reread that problem and make sure that your answer really makes sense. Some of the multiple-choice answers given as possible answers are really **steps** to the final solution. Make sure the answer you choose is really the **final** solution to the problem.

2. Take your time. You have plenty of time to finish the test. Read carefully.

3. Underline. You are permitted to write on your test booklet. To help keep yourself focused on the important parts of the problem, keep a pencil in your hand and underline. You might underline while you are reading, or you might read the entire problem, then reread and underline information that will help you answer the question. This technique helps you identify key information.

4. Focus on the first and last sentences of the word problems. First, make sure you understand the application, then use your mathematics skills to solve the problem. Finally, double check to be sure your answer **is** the final answer to the problem.

Using This Study Guide

The lessons in this study guide are no different than the test itself. Make sure that you have a working idea of the **math vocabulary.** **Memorize the math facts** and **be able to do the math operations** with or without a calculator that are outlined in each section. You will be given a math reference sheet to use during the test. This sheet will give you a list of formulas. Don't depend on this reference sheet to have <u>all</u> the formulas and facts that you will need for the test. You must know the formulas listed in each lesson and how to solve for all parts of each formula.

The **Guided Practice** reviews the basic information of the lesson and walks you through the solution process. The **Independent Practice** allows you to solve problems on your own. The **Practice Test** gives problems with multiple-choice, short answer and extended response answers just like the actual test. When answering the extended response questions, be clear in your wording and don't skip any steps in your thinking. If it takes four steps to work a problem, write them all down and mention the important math words that apply. If in doubt, write sentences <u>and</u> show your math work.

Sure, you will use this study guide to focus on areas where you need help, but also use it to double check those areas that you think you know. You have not used some of these mathematics concepts for several years. Ask for help from your teacher if you are unclear in any area. Reviewing <u>every</u> section may be just what you need to pass this test.

I wish you much success.

Sincerely,

Shari Wolf

Meeting the Challenge...Mathematics

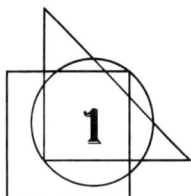

EQUIVALENT NUMBERS

Represent and use real numbers (decimals, fractions, percents, integers, exponential numbers, scientific notation) in a variety of equivalent forms.

FOCUS: We use real numbers every day. Can you solve these two real life problems?
(Look for the answers at the end of every lesson under Closure.)

1. Trisha needs $\frac{2}{3}$ cup of flour for her favorite cookie recipe. She has $\frac{1}{2}$ cup of flour left in the cupboard. Does she have enough flour for the recipe? The answer to this problem depends on the relationship of $\frac{2}{3}$ and $\frac{1}{2}$ and how to convert them to equivalent fractions.

2. Take a look at how a store can put an item "on sale." The BizMart has winter coats on sale at "40% off the regular price", but jeans are on sale at "$\frac{1}{4}$ off the regular price." Which sale is better? How can math help us decide?

PURPOSE: When you have completed this lesson, you will be able to convert a variety of real numbers in different forms into a common form (probably decimal) so that you can either compare them using $\neq, =, <, >, \leq, \geq$ or put them in ascending or descending order.

WHAT YOU NEED TO KNOW

MATHEMATICS AND PROBLEM SOLVING VOCABULARY:

1. Ascending order
2. Convert
3. Denominator of a fraction
4. Descending order
5. Equivalent numbers
6. Exponential numbers
7. Irrational numbers
8. Numerator of a fraction
9. Rational numbers
10. Real numbers
11. Scientific notation

Do you need to review these words? Use the Glossary in the back of the book.

MEMORIZE THESE FACTS:

1. Use $\pi = 3.14$ or $\frac{22}{7}$
2. Know these symbols:
   a. $=$, equal
   b. $\neq$, not equal
   c. $\geq$, greater than or equal to
   d. $\leq$, less than or equal to
   e. $>$, greater than
   f. $<$, less than

© 2003 Orange Frazer Press. All Rights Reserved.

Meeting the Challenge...Mathematics

BE ABLE TO PERFORM THESE OPERATIONS:

Convert

1. **Convert several fractions to equivalent fractions for comparison purposes.**
   Find the lowest common denominator of these fractions, then convert these fractions to equivalent fractions with this denominator.
2. **Convert a fraction to a decimal.**
   Divide the numerator by the denominator.
3. **Convert a percent to a decimal.**
   Move the decimal point two decimal places to the left and drop the percent sign.
4. **Convert real numbers such as decimals, fractions, percents, square roots, π, exponential numbers, numbers in scientific notation, and positive and negative numbers to an equivalent form.**
   Most of the time you will need to convert these real numbers to a decimal.

Compare and Order

1. Compare all types of real numbers using the following: $=, \neq, \geq, \leq, >, <$
2. **Order a list of real numbers of more than one type.**
   Take any real number, convert it to a decimal and then order.
3. **Determine if a set of real numbers is equivalent.**
   $\frac{3}{4} = .75 = 75\%$ (equivalent), $\frac{2}{3} \neq .75$ (not equivalent).
4. **Put a variety of positive and negative real numbers including irrational and exponential numbers, square roots and π in ascending order, descending order, or place on a number line.** Convert real numbers to decimals, but pay special attention to whether a number is positive or negative. Also, remember that if you are ordering negative numbers, the largest looking negative number is actually the smallest in value. For example, compare -7 and -3. The largest looking number, -7, is less than -3. The -7 would appear on the number line to the left (less than) of -3.

Evaluate

1. **Evaluate the exponential number.** Consider 3^4 where 3 is the base and 4 is the exponent. To evaluate this term, take the 3 (the base) times itself 4 times (the exponent) so that you are doing this: $3 \times 3 \times 3 \times 3 = 81$
2. **Evaluate scientific notation numbers.** The key concept is in moving the decimal point.
   a. If the exponent on the 10 is positive, move the decimal point that many places in the positive direction (to the right). For example, to evaluate 7.653×10^2, move the decimal point in 7.653 two places to the right so that 765.3 is the value of the expression.
   b. If the exponent on the 10 is negative, move the decimal point that many places in the negative direction (to the left). For example, to evaluate 4.37×10^{-3}, move the decimal point in 4.37 three places to the left so 0.00437 is the value of the expression. Yes, we added zeros.
   c. Sometimes zeros must be inserted as the decimal point is moved to the left or to the right.

BE ABLE TO PERFORM THESE CALCULATOR OPERATIONS:

1. Use the square root key. $\boxed{\sqrt{x}}$ 2. Use the square key. $\boxed{x^2}$
3. Convert a fraction to a decimal. Numerator ÷ Denominator

Meeting the Challenge...Mathematics

GUIDED PRACTICE

A) Convert to an equivalent form. Compare the numbers using the following: $<, >, \geq, \leq$
(Hint ~ When comparing two fractions, use equivalent fractions. Otherwise, convert the numbers to decimals and then compare.)

1) $\dfrac{2}{3}$ ☐ $\dfrac{5}{6}$

2) 0.4 ☐ 7% or 40%

3) 2^5 ☐ 38.2

4) $\dfrac{5}{8}$ ☐ 70%

5) 164,000 or 2000 ☐ 1.64×10^3

6) 7.4×10^{-2} ☐ 0.04

B) Arrange:
(Hint ~ Convert to decimals.)

7) Put in ascending order: $75\%, \dfrac{2}{5}, 0.17, 6.93 \times 10^{-1}$ _____

8) Put in descending order: $2.45 \times 10^2, 4^2, \pi, 2.50$ _____

9) Put in ascending order: $5, -3, \dfrac{2}{3}, -\dfrac{1}{2}$ _____

10) Place the letters on the number line below: $A = -3.5$, $B = -\sqrt{15}$, $C = -4$, $D = -3$

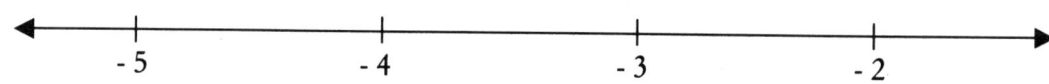

© 2003 Orange Frazer Press. All Rights Reserved.

Meeting the Challenge...Mathematics

INDEPENDENT PRACTICE

A) Tell whether the following pairs of real numbers are = or ≠

1) 0.7 ____ 7

2) $\dfrac{3}{5}$ ____ $\dfrac{7}{10}$

3) 0.5500 ____ 0.55

4) $\dfrac{3}{8}$ ____ 0.4

5) $\sqrt{169}$ ____ 13

6) 1.93×10^3 ____ 193000

B) Fill in the blanks using <, >, ≥, ≤

7) $\dfrac{5}{12}$ ____ $\dfrac{1}{3}$

8) $\dfrac{2}{3}$ ____ 75%

9) 55% ____ 0.06

10) $\dfrac{4}{5}$ or 0.2 ____ 0.85

11) 0.37 ____ 0.9

12) 4 ____ $3\dfrac{1}{2}$ or 1.5 or $\dfrac{8}{2}$

13) 2.8×10^{-2} ____ 0.04 or 0.028

14) 5000 ____ 4.954×10^3

C) Fill in the blanks using >, <, =

15) $\dfrac{2}{5}$ ____ $\dfrac{1}{2}$

16) 4 ____ $\sqrt{15}$

17) $\dfrac{8}{14}$ ____ $\dfrac{4}{7}$

18) 2 ____ $-1\dfrac{1}{2}$

19) 0.004 ____ 3.2×10^{-2}

20) $\sqrt{37}$ ____ 6

21) $\dfrac{7}{8}$ ____ $\dfrac{9}{11}$

22) $-1\dfrac{2}{3}$ ____ $-1\dfrac{1}{3}$

Meeting the Challenge...Mathematics

D) **Arrange in ascending order.**

23) $\frac{1}{3}$, $\frac{1}{5}$, $\frac{1}{4}$ _____

24) $6\frac{2}{3}$, 6.3, 6.4 _____

25) 0.45, 25%, $\frac{2}{3}$ _____

26) 0.75, $\frac{1}{6}$, 0.6 _____

E) **Arrange in descending order.**

27) 36.29, 36.209, 36.92 _____

28) −4.2, 0, −3.4, −0.8 _____

29) 0, π, −3, −9, 6, 4 _____

30) 3.5×10^{-2}, 2.07, 0.01, 4.1×10^{-3} _____

F) **Apply number comparisons and equivalent numbers.**

31)
```
   A   B
←──•───•───┼───┼───→
 -3  -2  -1   0   1
```
Which point is closer to $-2\frac{1}{2}$?

A   or   B   (Circle one)

32) If $x > -4.3$, give the letters for two possible values for x. _____ and _____ .

```
  D    C    B    A  F    E
←─•────•────•────•──•────•→
 -6   -5   -4   -3  -2   -1   0
```

© 2003 Orange Frazer Press. All Rights Reserved.

33) A farmer divided his land into four lots for new home construction. The lots were divided as follows: Lot A = $2\frac{1}{2}$ acres, Lot B = 2.6 acres, Lot C = $2\frac{7}{8}$ acres, and Lot D = $2\frac{3}{4}$ acres.

   a) Which lot is the largest? _____   b) Which lot is the smallest? _____

   c) Put lots A through D in order from smallest to largest.

   _____

34) The TriCities High School girls basketball team won 12 out of 16 games while their rival, Henderson High School, won 14 out of 20 games.

   a) Which team had the better record? _____

   b) Which team(s) won more than half their games? _____

   c) Which team(s) won at least 75% of their games? _____

35) Tell which letter on the number line below is closest to the value of $-\sqrt{10}$. _____

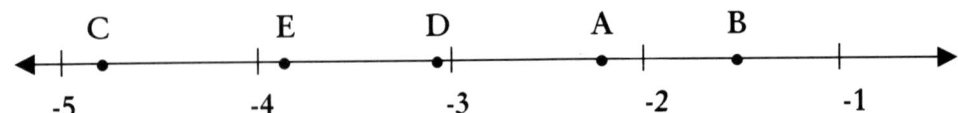

36) In his workshop, John is working with three pieces of sheet metal. The first sheet has thickness of 0.063 inches, the second has thickness of 0.015 inches, and the third has thickness of 0.032 inches. The blueprint says that he should be working with a piece of metal with a thickness between 2.4×10^{-2} and 1.4×10^{-2}.

   Which piece of metal should he use? _____

Meeting the Challenge...Mathematics

TEST PRACTICE

MULTIPLE-CHOICE QUESTIONS: Write the letter of the correct answer.

1) Which of the real numbers below lies between the numbers 3.1 and 3.2 ? _____

     a. 3.10   b. 3.21   c. 3.15   d. 3.019

2) Which of the following is a correct statement? _____

     a. $-\frac{2}{3} > -\frac{1}{2}$ b. $\frac{1}{2} < 0.5$ c. $0.25 > 10\%$ d. $\frac{3}{4} = 0.8$

3) Arrange the following in ascending order: 3^3, 2.532×10^1, $14\frac{1}{2}$, 2^4 _____

     a. 2^4, 3^3, 2.532×10^1, $14\frac{1}{2}$ b. 2^4, 2.532×10^1, $14\frac{1}{2}$, 3^3

     c. 3^3, 2.532×10^1, $14\frac{1}{2}$, 2^4 d. $14\frac{1}{2}$, 2^4, 2.532×10^1, 3^3

4) Fill in the blank. $\frac{7}{8}$ _____ 75 % _____

     a. <   b. >   c. ≥   d. ≤

5) Which value is closest to $\sqrt{19}$? _____

     a. 8.5   b. 4   c. 5   d. 3

SHORT ANSWER QUESTIONS:

6) The value of $-\sqrt{27}$ is between which two integers? _____ and _____

7) Tony needs a piece of lumber at least $89\frac{1}{4}$ inches long. In the garage, there is a piece of lumber that is $89\frac{5}{8}$ inches long.

     Is this piece of lumber long enough? _____ Why? _____

EXTENDED RESPONSE QUESTIONS:

8) Josh scored 21 out of 25 questions correctly. Jason scored 9 out of 10 questions correctly. Tell who had the better score and tell how you can use equivalent numbers to solve this problem.

Meeting the Challenge...Mathematics

9) Three pieces of rope have lengths of $3\frac{1}{2}$ feet, 3.75 feet and $3\frac{5}{12}$ feet. Explain how to put the three pieces of rope in ascending order. Tell which length is the longest piece of rope and show your work in the box below.

Answer = _____

10) John wants to buy $4\frac{11}{16}$ pounds of hamburger. All the packages of hamburger show their weights using decimals. List the steps he should take to choose the package he needs?

_____

_____

CLOSURE

Here are the solutions to the real life problems on page 1.

Problem #1 Trisha needs $\frac{2}{3}$ cup of flour and knows that she has $\frac{1}{2}$ cup left in the house. Does she have enough? Now what? We need to decide which fraction is larger, $\frac{1}{2}$ or $\frac{2}{3}$. If $\frac{2}{3}$ is larger, then she does not have enough flour. If $\frac{1}{2}$ is larger, then she has enough flour and can start cooking. Now compare. We need to convert both of these fractions to fractions with common denominators. In this case, six would be the common denominator. $\frac{2}{3} = \frac{4}{6}$ and $\frac{1}{2} = \frac{3}{6}$
We can see that $\frac{2}{3}$ ($\frac{4}{6}$ for comparing) is definitely larger than $\frac{1}{2}$ ($\frac{3}{6}$) which means that in our situation, Trisha should plan to go to the store because she does not have enough flour.

Problem #2 How can we tell which sale is better? First, we need to be able to compare a "40% off" sale with a "$\frac{1}{4}$ off" sale. In other words, we need to convert 40% and $\frac{1}{4}$ to equivalent numbers.
Convert both numbers to decimals and compare. 40% = 0.40 and $\frac{1}{4}$ = 0.25
The "40% off" sale is definitely the better sale. Call your friend and go shopping!

Meeting the Challenge...Mathematics

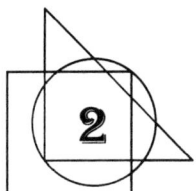

COMPUTING AND ESTIMATING

Compute and estimate with real numbers.

FOCUS: Let's use mathematics to help Trish with her checkbook.

Last Saturday, Trish ran errands to pay bills. Her first stop was at the Midtown Bank to deposit her paycheck. She needed to make sure that she had enough money in her checking account to cover her bills.

1. How does depositing her paycheck affect her checking account balance?

2. She will be paying her bills by writing checks from her account. How does each check written affect her account balance?

3. By paying the heating bill of $50 before the 10th of the month, Trish will receive a 5% discount. She paid the bill on the 4th of the month.
    - Give the amount of the heating bill check.
    - Show the way you would put the numbers and operations into the calculator the figure the discounted heating bill.

4. She found her optometrist bill for $28.50 that she did not pay last month. She must include a service charge of 2% on the balance.
    - Give the amount of this check.
    - List the steps for finding this amount using the calculator.

PURPOSE: When you have completed this lesson, you will be able to compute and estimate using real numbers to solve real world application problems.

WHAT YOU NEED TO KNOW

MATHEMATICS AND PROBLEM SOLVING VOCABULARY:

1. Key words and phrases for the following:
    a. Addition – sum, increased by, plus
    b. Subtraction – difference, decreased by, minus
    c. Multiplication – product, times, the word "of"
    d. Division – quotient, goes into, cut into parts

2. Key words for money-related applications:
    a. Annual, biannual, quarterly, semiannual
    b. Down Payment
    c. Finance charge
    d. Gross pay
    e. Interest
    f. List price
    g. Loan principal
    h. Net pay or take-home pay
    i. Overtime hours
    j. Overtime rate
    k. Payroll deductions
    l. Profit
    m. Reciprocal
    n. Sale price
    o. Service charge

Meeting the Challenge...Mathematics

MEMORIZE THESE FACTS AND FORMULAS:

1. Business and pricing formulas:
   a. Unit price = total price ÷ number of units
   b. Sales tax = total price × sales tax %
   c. Profit or loss = Gross income − expenses
      If the answer is positive, there is a profit. If negative, there is a loss.
2. Personal finance and payroll formulas and information:
   a. Gross pay = pay before deductions, or hours worked × hourly rate, or regular pay + overtime pay
   b. Overtime rate = 1.5 × regular rate also known as "time and a half"
   c. Overtime pay = hours worked over 40 × overtime pay rate
   d. Payroll deductions such as taxes, health care, savings, etc. are subtracted from your check.
   e. Net (take-home) pay = gross pay − deductions
   f. Checkbook math: Add deposits and interest.
      Subtract checks written, debit card deductions and checking account service charges.
   g. Interest = principal × rate × time (in years)
   h. Loan amount = principal + interest
3. Use $\pi = 3.14$ or $\frac{22}{7}$

BE ABLE TO PERFORM THESE OPERATIONS:

Fractions

1. **Add fractions.** Convert fractions to common denominators, add, then simplify.
2. **Subtract fractions.** Convert fractions to common denominators, subtract, then simplify.
3. **Multiply fractions.** Rename mixed numbers as improper fractions. Divide numerators and denominators by any common factors. Put the final answer in simplest form.
4. **Divide fractions.** Rename mixed numbers as improper fractions. Take the reciprocal of the divisor, and multiply the fractions. Divide the numerators and denominators by any common factors. Put the final answer in simplest form.

Decimals

1. **Add and subtract decimal numbers.** Line up the decimal points and then add or subtract.
2. **Multiply decimal numbers.** Multiply the numbers. The number of decimal places in the answer is the sum of the decimal places in the numbers multiplied. Zeros may need to be added to provide enough decimal places.
3. **Divide decimal numbers.** In the divisor, move the decimal point to the right as many places needed to convert the decimal to a whole number. Move the decimal point in the dividend the same number of places. Write zeros in the dividend, if necessary, to move the decimal point.

Estimation

1. **Estimate using whole numbers and decimals.** Estimate or round the number to the nearest whole number place values (10s, 100s, 1000s, etc.) or to decimal values [0.1 (tenths), 0.01 (hundredths), 0.001 (thousandths), etc.].
2. **Estimate money values.** Round off money values to the nearest $100, $10, or $1 or to the nearest cent. Make sure your answer is reasonable.

Meeting the Challenge...Mathematics

Business Applications

1. **Solve for the unit price.** Unit price = total price ÷ number of units
2. **Compare prices and find the best buy.** Solve for unit prices, then compare.
3. **Calculate the amount of change and tell the best combination of bills and coins to use.**
   Amount paid − total bill = change
   Make change using the least number of coins and bills possible.
4. **Find the profit on the sale of an item.** Profit = selling price of item − cost of item

Percent and Absolute Value

1. **Find what percent one number is of another.**
   Example: 12 is what percent of 16?
   Set up the equation: $12 = ?\% \times 16$ Calculate: $12 \div 16$ which is 0.75
   Then convert the decimal to a percent.   **Answer:** 75%   See Lesson #3 for the ratio method.
2. **Find a number when a percent of it is known.**
   Example: 25% of a number is 20, find the number.
   Set up the equation: $25\% \times ? = 20$ Calculate: $20 \div 0.25$ (25% is 0.25)
   Answer: 80   See Lesson #3 for the ratio method.
3. **Find the percent of a number.**
   Example: 30% of 90 is what number?
   Set up the equation: $30\% \times 90 = ?$ Convert: 30% to 0.30   Calculate: 0.30×90
   Answer: 27   See Lesson #3 for the ratio method.
4. **Evaluate absolute value.** Use absolute value to find the distance between points on a number line. First, subtract the two number values and then make the answer positive.
   Example: Find the distance from -5 to -1.
   Set up the absolute value: $|-5 - (-1)| = |-5 + 1| = |-4|$ **Answer:** 4

BE ABLE TO PERFORM THESE CALCULATOR OPERATIONS:

1. Use the $\boxed{\%}$ key to solve for all three parts of a percent problem.
   Most scientific calculators have a percent key. If yours does not, convert the percent to a decimal by moving the decimal point two places to the left. Use the decimal value for your calculations.
   a. Find the percent of a number. Example: Find 15.6% of 65
      Multiply the percent times the base number: $15.6\% \times 65$
      Calculate: 0.156×65 Answer: 10.14
   b. Find a number when a percent of it is known. Example: 45% of what number is 27?
      Set up the equation: $45\% \times ? = 27$ Calculate: $27 \div 45\%$ (or 0.45)   Answer: 60
   c. Find what percent one number is of another. Example: 2.4 is what percent of 1.6?
      Set up the equation: $2.4 = ?\% \times 1.6$ Calculate: $2.4 \div 1.6$ Result: 1.5
      Convert 1.5 to a percent.   Answer: 150%
2. Use the square root key.   Enter the number and push $\boxed{\sqrt{x}}$.

3. Use the square key.   Enter the number and push $\boxed{x^2}$.

© 2003 Orange Frazer Press. All Rights Reserved.

GUIDED PRACTICE

A) Solve and put the answer in simplest form.

1) $\dfrac{3}{4} + \dfrac{5}{6} =$ _____ = _____

2) $\dfrac{7}{8} - \dfrac{2}{3} =$ _____ = _____

3) $\dfrac{4}{5} \times \dfrac{15}{16} =$ _____ = _____

4) $\dfrac{7}{8} \div \dfrac{3}{4} =$ _____ = _____

5) $6\dfrac{1}{4} + 3\dfrac{2}{3} =$ _____ = _____

6) $7 - 2\dfrac{5}{8} =$ _____ = _____

7) $7\dfrac{3}{8} + 2\dfrac{5}{6} =$ _____ = _____

8) $6\dfrac{1}{2} - 3\dfrac{2}{3} =$ _____ = _____

9) $4\dfrac{5}{6} \times 4 =$ _____ = _____

10) $3\dfrac{1}{3} \times 3\dfrac{4}{5} =$ _____ = _____

11) $1\dfrac{7}{8} \div 3 =$ _____ = _____

12) $5\dfrac{1}{4} \div 8\dfrac{1}{6} =$ _____ = _____

B) Use an equation to solve the following percent problems:

13) What percent of 60 is 24?

   Equation: _____   Solution: _____

14) 15% of a number is 6. Find the number.

   Equation: _____   Solution: _____

15) 28% of 165 is what number?

   Equation: _____   Solution: _____

Meeting the Challenge...Mathematics

C) **Solve the application problems:**

16) The Wright Flyers are playing the Monroe Wolf Pack in the biggest game of the high school season. The Wright High School stadium has 50 rows of seats with 35 seats in each row. The ticket price is $5.25 and the game is a sellout.

    Find the total ticket sales. _____

17) Jack has been saving his money for the down payment on his first car. He needs $750.00 and has saved $516.00 already. He earns $54.25 a week from his part time job.

    a) How many more weeks will it take until he has the down payment saved? _____

    b) Show how to use estimation to verify this answer.
    _____

18) The Smither Corporation showed a profit of $7,600,000 for the 4th quarter of last year. This represents a gain of $1,200,000 over the 3rd quarter.

    a) What was third quarter's profit? _____

    b) What was the total profit for the two quarters? _____

D) **Use estimation to answer the following:**

19) Steve has $25.00 to spend at the electronics store. Use estimation to decide if he has enough money to buy a $14.75 cd and a $6.90 game.

    a) What estimation amounts did you use for the cd and game? _____ and _____

    b) Does he have enough money for both? _____

20) Judy spent $5.58 and gave the clerk a $20 bill. Use estimation and the rules for making change to decide if she should receive a $5 bill in change.   YES   NO   (circle one)

    Give your reasoning. _____

Meeting the Challenge...Mathematics

INDEPENDENT PRACTICE

Solve the following application problems rounding money answers to the nearest cent.

1) A 14-ounce box of snack crackers costs $1.79. A 24-ounce box of saltines costs $2.89.
   a) Which is the better buy? _____

   b) Why? _____

2) A chair sells for a cash price of $441.30. The chair could also be purchased by making a 10% down payment, and payments of $28.93 a month for 18 months.
   a) Give the down payment. $ _____
   b) Find the total price using credit. $ _____
   c) Calculate the finance charge. $ _____

3) The price of propane gas has increased 28% over last year. The Nelson family spent $650 on propane last year.
   a) What can they expect to spend this year? $ _____
   b) Find the monthly payment if this bill is paid in 12 equal installments. $ _____

4) Joyce worked $7\frac{1}{2}$ hours on Monday, $7\frac{3}{4}$ hours on Wednesday, and 7.25 hours on Friday.
   a) Which day did Joyce work the most hours?   _____
   b) Find the total hours she worked for the week.   _____
   c) Joyce earns $7.58 per hour. Find her earnings for the week. $ _____

5) George is going to build 5 shelves in his closet. Each shelf will be $8\frac{3}{4}$ inches wide.
   a) Can George cut the shelves from one sheet of plywood 48 inches wide?

        YES   NO   (circle one)

   b) Support your answer by providing an explanation.

      _____

48 in.

Meeting the Challenge...Mathematics

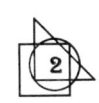

6) Jeanne bought two pairs of shoes. One pair cost $24.55 and the other $37.45. The sales tax rate was 6.5%.

   a) Find the sales tax. _____
   b) Find the total bill. _____
   c) Jeanne gave the clerk $100. How much change should she have received? _____
   d) Tell which bills and coins should be given as change.

   ___ $20 bills   ___ $10 bills   ___ $5 bills   ___ $1 bills

   ___ quarters   ___ dimes   ___ nickels   ___ pennies

7) Mrs. Larkin's class sold $2650.00 worth of candy for the school's annual candy sale. Jessie, a student in the class, sold $238.50 worth of candy. What percent of the class sales were Jessie's? _____

8) The grocery store has 492 cases of laundry soap, 258 cases of dishwasher soap, 134 cases of bath soap and 345 cases of liquid bath soap in stock. Estimate the total number of cases to the nearest ten.

   Laundry Soap = _____   Bath Soap = _____

   Dishwasher Soap = _____   Liquid Bath Soap = _____   Total = _____

9) Solve the following fraction problems. (Show all your work.)

   a) $11\frac{3}{8} + 15\frac{3}{4} =$ _____ = _____

   b) $\frac{5}{6} - \frac{3}{4} =$ _____ = _____

   c) $\frac{3}{10} \times \frac{8}{9} =$ _____ = _____

   d) $1\frac{3}{4} \times 3\frac{1}{7} =$ _____ = _____

   e) $1\frac{3}{4} \div 4\frac{3}{8} =$ _____ = _____

   f) $9\frac{1}{6} - 3\frac{5}{12} =$ _____ = _____

10) Steve worked 53 hours this week. His regular hourly wage is $9.50 per hour.

    a) Tell how many overtime hours he worked. _____

    b) Calculate his overtime rate using time and a half. _____

    c) List the steps used to find Steve's gross pay.

    _____

    d) Find his gross pay for the week. _____

Meeting the Challenge...Mathematics

11) The junior high athletic boosters club paid $519.75 for 275 sub sandwiches to sell for a moneymaking project. The subs were sold for $3.50 each.

Calculate the club's profit. _____

12) John had $738.14 in his checkbook on February 1. The following transactions took place in February: checks written for $225.65, $88.19 and $125.15, a deposit for $133.96, a service charge of $5.00

Find John's checkbook balance as of March 1. _____

TEST PRACTICE

MULTIPLE-CHOICE QUESTIONS: Write the letter of the correct answer.

1) Danny must nail some boards over a broken window $11\frac{1}{2}$ feet wide. He is using boards that are $2\frac{2}{3}$ feet wide. Which mathematics operation would you use to determine how many boards he needs? _____

    a. Addition   b. Subtraction   c. Multiplication   d. Division

2) Refer back to problem #1. How many boards does he need to completely cover the window? _____

    a. 3   b. 4   c. 5   d. 6

3) Mark bought a pair of hiking boots that were on sale for "$\frac{1}{4}$ off" the $130.36 list price. What was the sale price? _____

    a. $125.00   b. $32.59   c. $97.77   d. $65.18

4) At which price will one apple cost the least? _____

    a. 3 for $1.29   b. 5 for $1.85   c. 6 for $2.28   d. 2 for $0.95

5) A $4\frac{2}{3}$ ounce candy bar contains 5 grams of fat per ounce. How many grams are contained in the entire bar? Choose the closest answer. _____

    a. 20   b. 25   c. 21   d. 23

Meeting the Challenge...Mathematics

SHORT ANSWER QUESTIONS:

6) The cash price of Martin's new car is $7,500. He will make a down payment of $500 and finance the balance at an interest rate of 9% for 3 years.

   Find the following answers rounded to the nearest cent.

   a) Loan principal = _____

   b) Total interest for 3 years = _____

   c) Monthly payment = _____

   d) Total cost of the car = _____

7) When Terry started her diet two months ago, she weighed 154.75 pounds. She lost $7\frac{1}{4}$ pounds the first month and 4.2 pounds the second month. Her goal weight is 135 pounds.

   How many more pounds does she need to lose? _____

8) The local electronics store paid $231.00 for 21 game cartridges, then sold them for $23.95 each.

   a) How much profit did the store make on each cartridge? _____

   b) Did they make more than 100% profit on the cartridges?   YES   or   NO

EXTENDED REPONSE QUESTIONS:

9) At the flea market, new cds sell for $\frac{1}{2}$ their list price. Used cds sell for $\frac{1}{3}$ their list price.

   a) Show the math calculation you would use to figure the price of a new cd at the flea market.

      _____

   b) The list price of a new cd is $12.98. A new cd at the flea market would cost _____.

   c) The list price of a used cd is $14.79. A used cd at the flea market would cost _____.

Meeting the Challenge...Mathematics

10) Sue's annual car insurance costs include: $338.00 for bodily-injury liability, $192.00 for property damage, and $377.00 for collision coverage.

   a) Define semi-annual _____

   b) How much is her semi-annual premium? _____

   c) Explain the steps in finding the answer to this problem. _____
   _____

   d) Support your answer to Part c. Show your work in the box below.

   ```
                                                              Answer: _____
   ```

CLOSURE

Here are the solutions to the real life problem on page 9.

Mathematics and personal finances work hand in hand. Paying monthly bills on time and keeping accurate checkbook records helps you to avoid many financial problems. Trish knows that her checkbook balance must be large enough to cover her monthly bills at all times.

1) Depositing her paycheck adds to her checking account balance.

2) The amount of each check written is subtracted from the checkbook balance.

3) It's common practice to receive a discount for paying bills promptly. To find the discount on Trish's heating bill, take $50.00 (the heating bill) × 5% (the discount rate) = $2.50 (the discount). The heating bill check should be written for $47.50 ($50.00 − $2.50).

4) Trish forgot to pay her optometrist bill last month. A service charge should be added to this bill. To figure the service charge, take $28.50 (balance) × 2% (service charge) = $0.57 (service charge). The check should be written for $29.07 ($28.50 + $0.57).

Meeting the Challenge...Mathematics

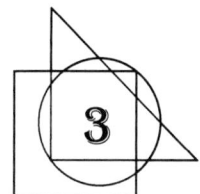

RATIO AND PROPORTION

Apply rates, ratios, proportions and percents.

FOCUS: Let's see where rates, ratios, proportions and percents could appear in our daily lives.

1. Vicki has budgeted $70 to buy a new winter coat. Does she have enough money to buy a coat regularly priced at $89 that is now marked down 30%? Use the sales tax at 7%.

2. Joe asked his boss to advance him part of his weekly pay that he would normally receive on Friday. On Tuesday Joe's boss gave him $65, which represented 40% of his weekly pay. What was Joe's total pay? How much money will Joe receive on Friday?

3. If Stacy can buy 2 candles for $0.79, how many candles can she buy with $3.00? Use a proportion to solve this problem.

PURPOSE: At the end of this lesson, you will be able to use rates, ratios, proportions, and percents to solve real world problems. Consumer applications (i.e., discounts, interest, tips, sale price, multiple markdowns), scale drawings, the distance formula ($d = r \times t$), and common units of measure (i.e., length, capacity, weight, time, and metric) are types of real world problems you will be able to solve.

WHAT YOU NEED TO KNOW

MATHEMATICS AND PROBLEM SOLVING VOCABULARY:

1. Cross multiply
2. Discount amount
3. Discount rate
4. Equivalent ratios
5. Finance charge
6. Gratuity (tip)
7. Installment loan
8. Interest amount
9. Interest rate
10. Means and extremes
11. Proportion
12. Rates
13. Ratio
14. Scale drawing

MEMORIZE THESE FACTS AND FORMULAS:

1. Distance = rate × time
2. Interest amount = principal × rate × time (in years)
3. Sales tax = total price × sales tax rate
4. Discount amount = total price × discount rate
5. Percent of decrease = Amount of decrease ÷ Original number
6. Percent of increase = Amount of increase ÷ Original number

© 2003 Orange Frazer Press. All Rights Reserved.

Meeting the Challenge...Mathematics

BE ABLE TO PERFORM THESE OPERATIONS:

Ratios

1. **Set up a ratio.** Compare two quantities by using a fraction bar, a colon, or the word *to*.
   Example: Write the ratio, 3 computers for every 7 students, in three ways.
   Answer: $\frac{3}{7}$ using a fraction, 3 : 7 using a colon, 3 to 7 using the word *to*

2. **Convert a ratio to common units.** Convert numbers with different units to the same unit.
   Example: Write the ratio, 2 feet to 3 yards, in three ways.
   Answer: Since 3 yards = 9 feet, the ratio is $\frac{2}{9}$, 2 : 9 or 2 to 9

3. **Simplify a ratio.** Write the ratio as a fraction and simplify by reducing to lowest terms.
   Example: Write the ratio, 15 minutes to 105 minutes, and simplify.
   Answer: $\frac{15}{105}$ simplifies to $\frac{3}{21}$

4. **Write a rate as a ratio.** Write the rate as a fraction and simplify by reducing to lowest terms.
   Examples: a) 32 miles per gallon = $\frac{32 \text{ miles}}{1 \text{ gallon}}$ b) 20 feet every 5 seconds = $\frac{20 \text{ feet}}{5 \text{ seconds}} = \frac{4 \text{ feet}}{1 \text{ second}}$

5. **Determine if ratios are equivalent.** If two ratios have the same value, they are equivalent.
   Examples: a) $\frac{4}{6} = \frac{2}{3}$ They are equivalent.   b) $\frac{9}{12} \neq \frac{2}{3}$ They are *not* equivalent.

Proportions

1. **Write a proportion.** Write two equivalent ratios as a proportion in fraction or colon form.
   Example: 5 is to 9 as 15 is to 27   **Answer:** using colons 5 : 9 = 15 : 27, using fractions $\frac{5}{9} = \frac{15}{27}$

2. **Identify a proportion.** A proportion shows two equivalent ratios.
   Example: Is $\frac{2}{3} = \frac{6}{9}$ a proportion? If the product of the means (3 × 6) equals the product of the extremes (2 × 9), the expression is a proportion.   **Answer:** Yes, it is a proportion.

3. **Solve a proportion.** When three of the four terms in a proportion are known, set the product of the means equal to the product of the extremes, and solve for the missing term.
   Example: $\frac{x}{8} = \frac{15}{40}$ Solve: $15 \cdot 8 = 40 \cdot x$ $120 = 40x$ **Answer:** $x = 3$

4. **Use proportions to solve percentage problems.**
   a. Find the base number when a percent of it is known. Example: 80% of a number is 48.
      Set up the equation: $\frac{80}{100} \cdot x = 48$ Set up the proportion: $\frac{80}{100} = \frac{48}{x}$
      Solve: $48 \cdot 100 = 80 \cdot x$ $4800 = 80x$ **Answer:** $x = 60$

   b. Find what percent one number is of another. Example: What % of 40 is 32?
      Set up the equation: $\frac{x}{100} \cdot 40 = 32$ Set up the proportion: $\frac{x}{100} = \frac{32}{40}$
      Solve: $32 \cdot 100 = 40 \cdot x$ $3200 = 40x$ **Answer:** $x = 80$

   c. Find the percentage of a number. Example: 65% of 90 is what number?
      Set up the equation: $\frac{65}{100} \cdot 90 = x$ Set up the proportion: $\frac{65}{100} = \frac{x}{90}$
      Solve: $65 \cdot 90 = 100 \cdot x$ $5850 = 100x$ **Answer:** $x = 58.5$

Meeting the Challenge...Mathematics

Business and Real Life Applications

1. **Find the gratuity (tip).** Customers "tip" for good service at the current rate of 15%.
   Example: Find the tip for a $28.14 restaurant bill.
   Set up the equation: Tip = $28.14 × 15%   Calculate: $28.14 × 0.15   **Answer:** $4.22

2. **Find the elapsed time between two clock readings.**
   Example: John clocked in at 8:30 a.m. and out at 1:15 p.m. Tell how many hours he worked.
   Solution: John worked from 8:30 a.m. until noon (3.5 hours), then from noon
   until 1:15 p.m. (1 hr. 15 min. = 1.25 hours).   **Total hours worked:** 4.75 hours

3. **Solve for d, r, or t in the distance formula.** Use $d = r \cdot t$ when solving for distance.
   Use $r = \frac{d}{t}$ when solving for rate. Use $t = \frac{d}{r}$ when solving for time.

4. **Solve for interest, principal, rate or time.** Use $i = p \times r \times t$ (in years)
   Example: Find the interest on a loan of $2000 at 9% for 2 years
   Solution: $i = \$2000 \times 9\% \times 2$ Calculate: $2000 × 0.09 × 2   **Answer:** $360.00

5. **Use multiple representations of a percent in word problems.** When considering percents, the whole of anything is 100%.
   Example: A "15% off" sale price is the same as 85% of the original price. The two percents total 100%.

6. **Calculate multiple discounts.** Calculate the first discount, then the second discount.
   DO NOT combine the two as one single discount.
   Example: In February an $85 winter coat was marked down 25%, and in April it was marked down an additional 35% for clearance. Find the clearance price. Solution: First, find the coat's price after the first discount. Then apply the second discount to get the clearance price. DO NOT take 60% (25% + 35%) off the original price. Remember that 25% off the original price is the same as 75% of the original price. The price after the first discount: $85 × 75% = $63.75. To find the price after the second discount, use 65% (100% – 35%).   Calculate: $63.75 × 65%   **Answer:** $41.44

7. **Use common units of measure and convert within their system.**
   Know conversions for:
   a) length (inch, foot, yard, mile)
   b) capacity (ounce, cup, pint, quart, gallon)
   c) weight (ounce, pound, ton)
   d) time (second, minute, hour, day, week, year)
   e) the metric system (liter, meter, gram)

8. **Set up and use proportions to find missing values in scale drawings.** The scale gives the relationship between the scale drawing and the actual size of the object in ratio form (scale : actual size). Use the given scale as one ratio in the proportion. Construct the other ratio by placing a variable representing the missing dimension along with the scale dimension (scale : real size).
   Example: The blueprint shows the bedroom of a house as 4″ × 6″. The scale is 1 inch : 3 feet.
   What are the actual dimensions of the bedroom?   Set up the proportion: $\frac{1}{3} = \frac{4}{width} = \frac{6}{length}$
   Solve for width: $\frac{1}{3} = \frac{4}{w}$, $w = 12$ feet   Solve for length: $\frac{1}{3} = \frac{6}{l}$, $l = 18$ feet

Meeting the Challenge...Mathematics

BE ABLE TO PERFORM THESE CALCULATOR OPERATIONS:

Use the % key to solve for all three parts of a percent problem. Most scientific calculators have a percent key. If yours does not, convert the percent to a decimal by moving the decimal point two places to the left. Use the decimal value for your calculations.
   a. Find the percent of a number. Example: 18% of 70
      Multiply the percent times the base number: 18% × 70
      Calculate: 18% (or 0.18) × 70   Answer: 12.6
   b. Find a number when a percent of it is known. Example: 15% of what number is 3?
      Set up the equation: 15% × ? = 3
      Calculate: 3 ÷ 15% (or 0.15)   Answer: 20
   c. Find what percent one number is of another. Example: 32 is what percent of 128?
      Set up the equation: 32 = ? % × 128
      Calculate: 32 ÷ 128   Result: 0.25   Convert 0.25 to a percent.   Answer: 25%

GUIDED PRACTICE

A) There are 64 cats and 20 dogs housed with the local veterinarian.
    Find the following ratios in simplest form.

    1) cats to dogs = _____   2) dogs to both = _____

    3) dogs to cats = _____   4) cats to birds = _____

B) Tell whether the following are proportions.

    5) $\frac{4}{7} = \frac{3}{5}$ Yes No   6) $\frac{3}{9} = \frac{1}{3}$ Yes No   7) $\frac{2}{5} = \frac{6}{15}$ Yes No

C) Solve these proportions.

    8) $\frac{x}{5} = \frac{4}{10}$ 9) $\frac{8}{15} = \frac{36}{x}$ 10) $\frac{9.6}{x} = \frac{0.024}{3}$ 11) $\frac{4}{9} = \frac{5\frac{1}{3}}{x}$

    x = _____   x = _____   x = _____   x = _____

D) Write the rates as ratios. Simplify if possible.

    12) 2 apples for $0.95 _____   13) 3 doughnuts for $1.08 _____

Meeting the Challenge...Mathematics

E) **Write the proportions and solve.**

  14) If 2 slices of cheese contain 230 calories, how many calories would 5 slices contain?

  $\underline{} = \underline{}$

  Solve:

  Answer: _____

  15) If three candy bars cost $1.39, how many candy bars can be purchased with $2.50?

  $\underline{} = \underline{}$

  Solve:

  Answer: _____

  16) The scale of a drawing is 1 in. : 4 ft. The blueprint shows the living room with dimensions of 4 in. by 5 in. Set up two proportions to find the actual size of the living room.

  $\underline{} = \underline{}$ $\underline{} = \underline{}$

  The actual dimensions of the living room are _____ by _____.

F) **Set up the proportions and solve the following percent problems.**

  17) What % of 40 is 24?   18) 5% of 20 is what number?   19) 16% of what number is 12?

  $\underline{} = \underline{}$ $\underline{} = \underline{}$ $\underline{} = \underline{}$

  Solve: Solve: Solve:

  Answer: _____   Answer: _____   Answer: _____

INDEPENDENT PRACTICE

Solve the following application problems rounding money answers to the nearest cent and other answers to two decimal places.

1) Celia earned $50 on Monday. She spent $35 of that amount and saved the rest. In lowest terms, give the following ratios:

   a) amount spent to the amount earned _____

   b) amount saved to the amount earned _____

   c) amount spent to the amount saved _____

   d) amount saved to the amount spent _____

Meeting the Challenge...Mathematics

2) Lisa received $18.60 in tip money from 12 customers. At the same rate, how much tip money will Lisa receive from 20 customers? Show your work below.

Answer: _____

3) Ray traveled 220 miles and used 12.8 gallons of gasoline. Amber traveled 132 miles and used 7.2 gallons of gasoline.

   a) What proportion would you use to find Ray's gas mileage? _____ Amber's? _____

   b) Find Ray's gas mileage. _____

   c) Find Amber's gas mileage. _____

   d) Which person has the better gas mileage? _____

4) Solve for the missing value. Show your work.

   a) $\dfrac{5}{9} = \dfrac{40}{y}$

   b) $\dfrac{x}{32} = \dfrac{3}{8}$

   c) $\dfrac{3}{4} = \dfrac{b}{24}$

   Equation: _____   Equation: _____   Equation: _____

   Answer: _____   Answer: _____   Answer: _____

5) Write the proportions for the following problems:

   a) Ramon worked 8 hours and was paid $51. At that same rate, how much would he earn in 27 hours?

   ___ = ___

   b) Kara traveled 286 miles in 5.25 hours. At the same rate, how long would it take her to travel 390 miles?

   ___ = ___

   c) Anita delivers 21 newspapers in $\tfrac{3}{4}$ hr. At the same rate, how many newspapers can she deliver in 2 hours?

   ___ = ___

   d) Linda's chili recipe calls for $1\tfrac{1}{2}$ lbs. of hamburger and 3 tsp. chili powder. How much chili powder will she need for 4 lbs. of hamburger?

   ___ = ___

Meeting the Challenge...Mathematics

6) To buy her first car, Kelly needs to borrow $1450 from her parents for two years. She will pay 9% interest. Show your work in finding the following:

   a) Interest amount _____ $ _____

   b) Total loan amount _____ $ _____

   c) Monthly payment _____ $ _____

7) Find the missing values. Show your work.

   a) Jacket price $42.90   b) Stereo price $215.00
      Discount rate   25%   Sale price $172.00

      Discount amount $_____   Discount amount $_____

      Sale price $_____   Discount rate   _____%

8) As a special promotion, a bottle of aspirin is advertised containing 15% more aspirins for the same price.

   a) Explain how to find the number of aspirins in the sale bottle. _____
   _____
   _____

   b) Normally the aspirin bottle contains 220 aspirins. The number of aspirins in the sale bottle is_____. (Show your work.)

9) The Lopez family spent $65.00 for dinner. Mr. Lopez left a 15% tip.

   a) Compute the tip.   Tip = $_____

   b) Compute the total bill.   Total Bill = $_____

   c) How much change should he have received from $100.00?   Change = $_____

   d) Tell which bills and coins should be given as change.

      ____ $20 bills   ____ $10 bills   ____ $5 bills   ____ $1 bills
      ____ quarters   ____ dimes   ____ nickels   ____ pennies

© 2003 Orange Frazer Press. All Rights Reserved.

Meeting the Challenge...Mathematics

10) If the discount rate on an item is 40%, what percent of the original price is the sale price? _____

11) Kevin received a 5% raise. His hourly rate was $7.10. What is his new hourly rate? _____

12) Renee works on an assembly line. She can sew a zipper into a jacket every three minutes. On Tuesday she worked from 7:45 a.m. to 11:30 a.m. and from 12:15 p.m. to 4 p.m.

    a) How many hours did she work? _____ hours

    b) How many zippers did Renee sew into jackets during this time?
                        _____ zippers

13) A recipe for 3 dozen cookies calls for $3\frac{1}{4}$ c. flour.

    Write a proportion to find the flour needed for 5 dozen cookies.   ____ = ____

Use the scale drawing of Jan's condo to answer the following questions. Scale: 1 in. = 6 ft.

14) Find the missing scale dimensions.

    a = _____ in.
    b = _____ in.
    c = _____ in.

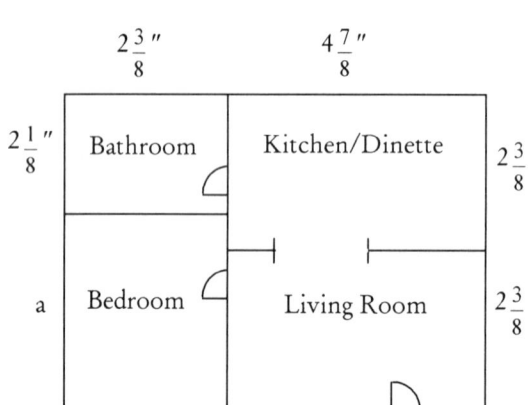

15) What are the actual dimensions of the following?

    a) kitchen/dinette   width = _____   length = _____
    b) living room   width = _____   length = _____
    c) bathroom   width = _____   length = _____
    d) bedroom   width = _____   length = _____

TEST PRACTICE

MULTIPLE–CHOICE QUESTIONS: Write the letter of the correct answer.

1) The store sells 3 videos to every 7 DVDs. What is the ratio of the number of videos sold to the total number of videos and DVDs sold? _____

    a. 3 to 7   b. 7 to 3   c. 3 to 10   d. 7 to 10

2) If a bag of mixed nuts has a ratio of 2 pounds of cashews to 3 pounds of peanuts, how many pounds of peanuts would be needed to make 20 pounds of the mix? _____

    a. 10   b. 8   c. 12   d. 15

Meeting the Challenge...Mathematics

3) T-shirts priced at $16.50 each are on sale 2 for $29.70. If two t-shirts are purchased, find the discount rate.

    a. 12.2%   b. 10%   c. 9%   d. 15%

4) Maria and Tom spent $350.00 for their wedding rehearsal dinner. Compute the tip at 20%.

    a. $70.00   b. $35.00   c. $42.50   d. $60.00

5) Howard's wage was raised from $6.25 to $6.50 per hour. What percent raise did he receive?

    a. 5%   b. 4%   c. 2%   d. 6%

SHORT ANSWER QUESTIONS:

6) An oil well can pump 485,239 barrels of oil in 3 days. Write a proportion to find the number of barrels of oil pumped at the same rate in 7 days.

    Proportion: ———— = ————

7) The ratio of overtime pay to regular pay is 3 : 2. If Chris earns $72 for a regular 8-hour day, what does she earn for 2.5 hours of overtime? Show the proportion and full solution.

    Proportion: ———— = ————   Answer: _____

EXTENDED RESPONSE QUESTIONS:

8) Joe can make 15 parts in 10 minutes. On Friday, he worked from 1:30 p.m. to 4:50 p.m. Find the total number of parts that he produced that day. Support your answer by providing an explanation. Show your work in the box below.

Answer: _____ parts

© 2003 Orange Frazer Press. All Rights Reserved.

Meeting the Challenge...Mathematics

9) The store charged $340.00 for a refrigerator.

   a) Explain how to find the store's cost of the refrigerator knowing that they made a 38% profit.

   _____

   b) Find the store's cost of the refrigerator.

   _____

10) On a scale drawing, $1\frac{1}{2}$ in. = $3\frac{1}{2}$ ft. Set up and solve a proportion to find the distance represented by $2\frac{1}{4}$ inches. Give your answer in fraction form.

    Proportion: ——— = ———

    Solve:

    Answer: _____ ft.

CLOSURE

Let's apply our knowledge of rates, ratios, proportions and percents to solve the problems on page 19.

Problem #1 Vicki must determine if the total cost of the coat is $70 or less. The total cost would be the sale price plus 7% sales tax.
Sale price = original price ($89) – discount ($89 × 30% = $26.70) = $62.30
Total price = sale price ($62.30) + sales tax ($62.30 × 7% = $4.36) = $66.66
Vicki can buy her new winter coat.

Problem #2 Joe has a nice boss that will give him some of his weekly pay before payday. Joe received $65, or 40% of his pay before payday. Set up an equation using: 40% of his weekly pay = $65
Equation: 0.40 × weekly pay = $65.00   Solve: Weekly pay = 65.00 ÷ 0.40 = $162.50
Joe will receive $97.50 ($162.50 – $65.00) on Friday.

Problem #3 Use a proportion to find how many candles Stacy can buy for $3.00. Set up the proportion with the number of candles in the numerator of the fraction and the cost of the candles in the denominator.
Proportion: $\frac{2}{0.79} = \frac{c}{\$3.00}$ Cross multiply: 0.79c = 6.00   Solve: c = 6.00 ÷ 0.79
Answer: 7.5949...   Stacy can buy 7 candles with money left over.

Meeting the Challenge...Mathematics

MATHEMATICAL PROCESSES
Communicating with Math

PURPOSE: In this lesson you will use mathematical language and notation to represent problem situations and express mathematical ideas clearly. You will also be able to locate, interpret and clearly communicate mathematical ideas, processes and solutions.

A) To determine the damages an insurance company should pay Mr. Molton to settle a lawsuit, the foreman of the jury had each jury member write what would be considered a fair settlement amount. The results were as follows: $1000; $1500; $1250; $2000; $1700; $2000; $50,000; $1650; $2000; $2250; $1800; $1400

   1) Find the measures of central tendency of this data.

   2) Explain what each measure means within the context of this problem.

   3) Discuss which measure of central tendency would most fairly represent the jury's opinion.

   4) Mr. Molton has $2000 in legal fees. Based on your decision for the settlement amount, did Mr. Molton receive enough money to pay his legal fees?

B) Use these six angles to answer the probability questions.

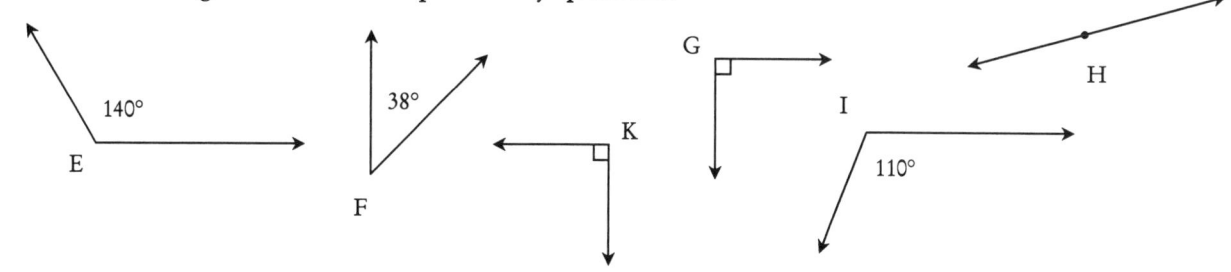

   1) What is the probability of randomly selecting an acute angle? _____
   2) What is the probability of randomly selecting an obtuse angle? _____
   3) What is the probability of randomly selecting an acute or straight angle? _____
   4) If two angles are chosen at random, what is the probability that both angles will be right angles? _____
   5) If two angles are chosen at random, what is the probability that one of them will be a right angle? _____

Meeting the Challenge...Mathematics

C) Use appropriate mathematics vocabulary words and expressions to describe these situations.

1) We can go to the pool, if Jamie has time to go with us. _____
2) I wish I knew how long the path is around Crater Lake. _____
3) I have 14 friends within 2 miles of my house. _____
4) A cheerleader with his/her arms extended is the _____ of a cartwheel.
5) I know the shortest path from the southeast corner to the northwest corner of my neighbor's rectangular cornfield. _____
6) That plan works every time. _____
7) You say it one way. I say it another way. We are both saying the same thing. _____
8) Being average means different things to different people. _____
9) John is walking to Natasha's house. As Kevin walks to Joe's house, he stops and talks with John as they cross paths. _____
10) How much is this object worth? _____
11) Whatever you decide doesn't matter to me. _____
12) I am not sure if I should go up or down, right or left. Where am I? _____
13) This object is a squashed square. _____
14) The Leaning Tower of Pisa is not _____ to the ground.

D) Now it's your turn to write an expression to describe these mathematics vocabulary words.

1) theoretical probability _____

2) similar figures _____

3) bisector _____

4) degree of an equation _____

5) regular polygon _____

6) surface area _____

7) reciprocal _____

8) proportion _____

Meeting the Challenge...Mathematics

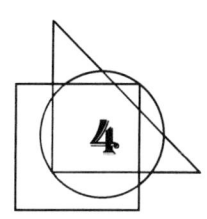

MEASUREMENT

4 Use measurement conversions and right triangle trigonometry in working with two-dimensional figures and three-dimensional objects.

FOCUS: Many areas of our everyday life include the use of measurement in both the U.S. customary system and in the metric system. Being able to work with both systems is an important skill.

In Canada, gasoline is sold by the liter and in the United States it is sold by the gallon. Discuss how you would compare the prices of gasoline in the U.S. and in Canada. Go to the Internet to get a current price for a liter of gasoline in Canada and compare prices. If you were going to Canada this weekend, should you fill your gas tank before crossing the border?

PURPOSE: When you have completed this lesson, you will be able to work with measurement conversions, appropriate units and indirect measurement through formulas.

WHAT YOU NEED TO KNOW

MATHEMATICS AND PROBLEM SOLVING VOCABULARY:

1. Central angle
2. Cosine
3. Major arc
4. Minor arc
5. Protractor
6. Sine
7. Tangent
8. Trigonometry

MEMORIZE THESE FACTS AND FORMULAS:

1. The trigonometric ratios (sine, cosine and tangent) are used to find missing sides and angles in a right triangle.

   $$\sin \angle A = \frac{opposite}{hypotenuse} \qquad \cos \angle A = \frac{adjacent}{hypotenuse} \qquad \tan \angle A = \frac{opposite}{adjacent}$$

   The sides of the right triangle to the right are labeled with respect to $\angle A$.

2. Use the Measurement Conversion Chart in the back of this study guide to convert between U.S. customary units and metric units.

3. Review the equivalencies within the U.S. customary measurement system and the metric system. Do not depend on a reference chart to provide facts such as there are 8 ounces in a cup or 2000 pounds in a ton. Memorize the most used equivalencies listed on the Measurement Study Sheet.

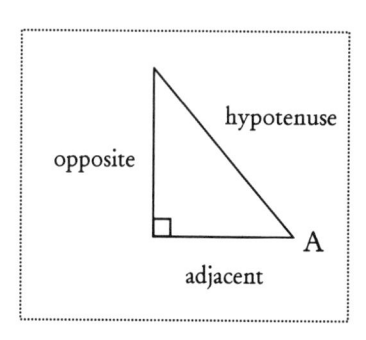

BE ABLE TO PERFORM THESE OPERATIONS:

Converting measures and rates

1. **Compare and order the relative size of common U.S. customary units and metric units.**

   Example 1: Which is larger, a meter or a yard?
   Solution: If you compare a yard stick to a meter stick, the meter stick is just a few inches longer. If you use a conversion chart, a meter is approximately 39 inches and a yard is 36 inches.
   Answer: a meter

   Example 2: Which is larger, 4 liters or 1 gallon?
   Solution: The conversion chart tells us that a liter is just a little larger than a quart. It takes 4 quarts to make a gallon.
   Answer: Since each liter is a littler larger than a quart, four liters is more than one gallon.

2. **Use formulas and proportions to convert from one system to another.**

   Example 1: Convert 75°F to °C.
   Solution: Use the conversion formula.
   $$C = \frac{5}{9}(F - 32)$$
   Substitute: $C = \frac{5}{9}(75 - 32)$ $C = \frac{5}{9}(43)$
   Answer: 23.9°C (rounded to one decimal place)

   Example 2: Convert 1700 grams to pounds.
   Solution: We know from the conversion chart that 453.59 grams are in a pound. So, this measure in grams is approximately 4 pounds. Calculate 1700 ÷ 453.59 to find the exact number of pounds.
   Answer: 3.75 pounds (rounded to two decimal places)

3. **Convert rates within the same measurement system.**

   Example: It takes Krista 15 minutes to drive 10 miles to school. Express her travel rate in miles per hour, feet per hour and feet per minute.
   Solution: Krista's travel rate is 10 miles/15 minutes (miles per minute).
       a) To convert the rate to miles/hour (miles per hour), we need to create an equivalent fraction with 1 hour (60 minutes) in the denominator. Multiply both the numerator and denominator by 4 (15 × 4 = 60 minutes).
   Answer: 40 miles/60 minutes = 40 miles/ hour
       b) To convert the answer from part a to feet/hour (feet per hour), convert 40 miles to feet. (40 × 5280 = 211,200)
   Answer: 211,200 feet/hour
       c) To convert the answer from part b to feet/minute (feet per minute), multiply the denominator by 60 to convert one hour to 60 minutes.
   Answer: 211,200 feet/60 minutes = 3520 feet/minute

Deriving and using formulas

1. **Derive formulas for surface area and volume for three-dimensional objects.**

   Example: Find the surface area of a pyramid with a square base of 6″ and an altitude of 4″.
   Solution: The surface area of a pyramid is made up of a square base and four triangular lateral faces. First find the area of the base (6″ × 6″ = 36 square inches) and then add this to the area of the triangular lateral faces $4[\frac{1}{2} \times base(6) \times slant\ height(5)]$. See Lesson #5 for step-by-step directions.
   Answer: surface area = 36 + 4(15) = 96 square inches

2. **Use the formulas for triangles, quadrilaterals and circles to find the surface area and volume of prisms, pyramids and cylinders and to find the volume of spheres and cones.**
   Go to Lesson #5 for examples.

Meeting the Challenge...Mathematics

Using a protractor and right triangle trigonometry

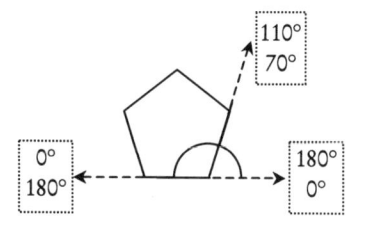

1. **Use a protractor to find the interior and exterior angles of a regular convex polygon.** This very small protractor gives the interior angles (110°) and the exterior angles (70°) of the pentagon. See Lesson #8 for additional methods to find these angle measures.

2. **Find missing sides and angles using right triangle trigonometry.** 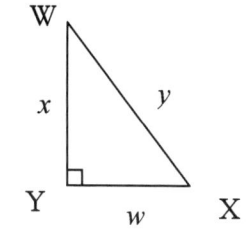
   Example 1: If m∠X = 50° and $w = 3.5''$, find m∠Y, m∠W, y and x.
   Solution a: ∠X and ∠W are complementary, so the sum of their measures is 90°.
   Answer a: m∠W = 40° (90° - 50°)
   Solution b: Solve for side x. With respect to ∠X, w is the adjacent side and x is the opposite side. The trigonometric function involving adjacent and opposite sides is the tangent function.
   $$\tan 50° = \frac{opposite}{adjacent} = \frac{x}{3.5}$$
   Use either your scientific calculator or look up the value to find the tan 50° = 1.192. Substitute the value into the equation. $1.192 = \frac{x}{3.5}$ Solve for x by cross multiplying.
   Answer b: $x = 4.2$ inches
   Solution c: Solve for side y. Since we have solved for x and w, we can use the Pythagorean theorem to solve for y.
   Answer c: $y = \sqrt{3.5^2 + 4.2^2} = \sqrt{29.89} = 5.5$ inches

3. **Use right triangle trigonometry to find an angle measure when two sides are known.** Determine which sides (opposite, adjacent or hypotenuse) are given in a right triangle. Choose a trigonometric function that includes those given sides, and find the ratio of those sides. Then use a trigonometric table or scientific calculator (2nd function) to find the angle measure.
   Example: With respect to ∠A in a right triangle, the opposite side is 17 and the hypotenuse is 21. Find m∠A.
   Solution: $\sin\angle A = \frac{17}{21} = 0.8095$ Use your scientific calculator or a trigonometric table to find m∠A.
   Answer: m∠A = 54.05°

BE ABLE TO PERFORM THESE CALCULATOR OPERATIONS:

1. Use your scientific calculator when working with trigonometric functions (sin, cos, and tan).
2. Use your calculator as needed when converting units.

GUIDED PRACTICE

A) Give the measurement unit from each system that best describes these lengths.

	U.S. Customary	Metric
1) the distance from Cleveland to Portsmouth		
2) the width of your school room		
3) the length of your ring finger		
4) the width of your thumb		

Meeting the Challenge...Mathematics

B) Tell which is larger.

5) centimeter or inch _____   6) gram or pound _____

7) liter or gallon _____   8) kilometer or mile _____

C) Use formulas and conversion facts to determine the following measures. Round to two decimal places.

9) Find °F if the temperature is 20°C. _____   10) Convert 4 inches to centimeters. _____

11) Convert 12 centimeters to inches. _____   12) Find °C if the temperature is 90°F. _____

13) Convert 6 liters to gallons. _____   14) Convert 75 grams to ounces. _____

INDEPENDENT PRACTICE

A) Convert these rates.

If Jean traveled 25 miles in 30 minutes, find these rates.

1) miles per hour _____   2) miles per minute _____   3) miles per second _____

4) feet per hour _____   5) feet per minute _____   6) feet per second _____

If Natasha bicycled 3 kilometers in 15 minutes, find these rates.

7) kilometers per hour _____   8) kilometers per minute _____   9) kilometers per second _____

10) meters per hour _____   11) meters per minute _____   12) meters per second _____

B) Describe how to find the surface area of the cylinder. Use sketches and be sure to include which 2-dimensional figures actually make up the object. Find the surface area.

13) cylinder: diameter = 10 feet, height = 25 feet

surface area _____

Meeting the Challenge...Mathematics

C) Use right triangle trigonometry to find the missing sides and angles.

14) m∠A = 55°   b = 7 in.

   a = _____   c = _____

15) m∠B = 32°   c = 12.4 ft.

   b = _____   a = _____

16) a = 14 cm.   b = 10 cm.

   c = _____   m∠A = _____

   m∠B = _____

17) a = 6.8 m   c = 15.2 m

   b = _____   m∠A = _____

   m∠B = _____

TEST PRACTICE

MULTIPLE-CHOICE QUESTIONS:

1) Rhonda is going to make six batches of a recipe that requires 3 cups of milk for each batch. She has a full quart in the refrigerator. What is the smallest amount of milk she can purchase to be sure she has enough for the six batches?

   a. 3 quarts   b. 1 gallon   c. 1 gallon, 1 quart   d. 2 gallons

2) Which lengths are in order from smallest to largest?

   a. meter, foot, kilometer
   b. inch, centimeter, meter
   c. yard, kilometer, mile
   d. inch, meter, yard

3) Which trigonometric function would you use to find side k?

   a. sine
   b. cosine
   c. tangent
   d. none of these

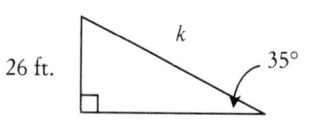

4) The average temperature for January last year was 38.3°F. In Celsius this temperature is

   a. 6°C   b. 3.5°C   c. 12.5°C   d. 10°C

5) A wheel makes 8 revolutions per second. How many revolutions will it make in an hour?

   a. 480   b. 28,800   c. 10,080   d. 25,008

SHORT ANSWER QUESTIONS:

6) A dripping faucet can waste about 100 liters of water a week. At $0.15 per gallon, what is the cost of the wasted water for six months?

cost of wasted water: $ _____

7) Look at the circle with a central angle of 60° and a radius of 9 inches.
   a) Describe the relationship between the size of the central angle and the length of the arcs defined by the sides of the central angle.

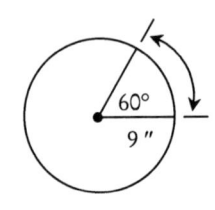

   b) Find the lengths of the major and minor arcs defined by the sides of the central angle.

   minor arc _____   major arc _____

EXTENDED RESPONSE QUESTIONS:

8) Mr. Thornburg is preparing to build an additional storage bin for his shelled corn. He wants to build a cylindrical bin large enough to hold 10,000 bushels of corn. There will be no storage in the cone-shaped cover. He is planning to construct the bin in a space 25 feet in diameter. Draw a sketch of the bin. Discuss the options that Mr. Thornburg has in the design of this bin including the size of the base and height. List other facts that must be known to solve this problem and complete the project.

9) Jordan and his family traveled to Canada for the weekend. Jordan noticed that the speed limit in Michigan was 70 m.p.h. When they entered Canada, Jordan noticed the speed limit became 80 k.p.h. (kilometers per hour). Explain the change in the speed limit. Describe the conversion process used to change miles per hour to kilometers per hour. Compare the two speed limits and discuss reasons for the differences.

10) Describe how to find the surface area of a pyramid with an 8 inch square base and an altitude of 12 feet. Be sure to include a sketch and tell which 2-dimensional figures actually make up the object. Find the surface area.

CLOSURE

The key to comparing the U.S. price of gasoline to the Canadian price is to know there are 3.785 liters in one gallon.
To compare the price of gasoline, multiply the Canadian price per liter × 3.785 which gives you a price to compare with the U.S. price per gallon.

Meeting the Challenge...Mathematics

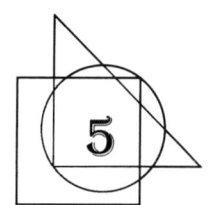

LENGTH, PERIMETER, AREA, SURFACE AREA AND VOLUME

Use measurement techniques in working with two-dimensional figures and three-dimensional objects.

FOCUS: Ohio's hot summers serve as a great incentive to purchase an above ground swimming pool. Geometry and measurement can be very useful in answering questions concerning a pool.

Brandon's new above ground family pool is in the shape of a circle 20 feet across and 4.5 feet deep. The following questions must be answered in order to keep the pool running efficiently and as economically as possible.
1. What would be the size of an insulated pool cover large enough to hang one foot over the rim of the pool?
2. The cost of the pool cover is $0.89 per square foot. What is the cost of the pool cover?
3. If the pool is filled to the 4-foot level, how many cubic feet of water will it hold?
4. The pool will be filled by a local water company. They charge $0.18 per cubic foot. What would be the cost to fill the pool?
5. A privacy fence will be installed around the pool. How many feet of fencing will be needed to circle the pool allowing for a 3-foot walkway between the pool and the fence? Draw a sketch and label it.

PURPOSE: When you have completed this lesson, you will be able to use a variety of measurement techniques including scale drawings, formulas and geometric relationships to find length, perimeter, area, surface area, and volume.

WHAT YOU NEED TO KNOW

Note: A formula sheet will be provided to all students during the OGT Mathematics Test. Many of the following length, area, and volume formulas will be included on the formula sheet, but your success on the test will depend on your knowledge of how to use these formulas. **Do not** be totally dependent on the formula sheet.
Keep the following thoughts in mind:
- Become **very** familiar with the formulas and when to use them.
- Know how to evaluate any formula for length, area and volume.
- Be able to solve for any missing variables in the formula.

MATHEMATICS AND PROBLEM SOLVING VOCABULARY:

1. Altitude
2. Circumference
3. Diagonal
4. Lateral area
5. Lateral edge
6. Lateral face
7. Perimeter
8. Pythagorean theorem
9. Semicircle
10. Slant height
11. Surface area
12. Volume

© 2003 Orange Frazer Press. All Rights Reserved.

MEMORIZE THESE FACTS AND BECOME FAMILIAR WITH THE FORMULAS:

1. In working with measurements, it is extremely important to know and use the proper units.
   - Units for length are in linear units such as inches, feet, miles, millimeters, meters, etc.
   - Units for area and surface area are in square units such as square inches, square meters, etc.
   - Units for volume are in cubic units such as cubic centimeters, cubic yards, etc.
2. The **perimeter** of a figure is the sum of the lengths of the sides or the distance around the outside.
3. Use these formulas to find length. They are included on the formula sheet at the back of this study guide.
   - Use the distance formula, $d = \sqrt{(x_2 - x_1)^2 + (y_2 - y_1)^2}$, to find the distance between two points on a coordinate graph. See Lesson #6 for more details.
   - Use the **Pythagorean theorem**, $c^2 = a^2 + b^2$, to find the lengths of the sides of a right triangle. In the formula, a and b are the legs (sides that form the right angle) and c is the hypotenuse (the longest side opposite the right angle).
   - Use the circumference formula, $C = \pi \cdot d$ (where d represents the diameter of the circle), to find the **circumference** (perimeter) of a circle.
4. Use the following formulas to find the areas of the two-dimensional figures. These formulas are also included on the formula sheet at the back of this study guide. Refer to this formula sheet just as you would during the actual test.

Two-dimensional Figures	Area Formula	Explanation
Triangle	$A = \frac{1}{2}bh$	The base (b) is the side to which the altitude or height (h) is drawn to perpendicular.
Parallelogram	$A = bh$	The base (b) is the side to which the altitude or height (h) is drawn to perpendicular.
Rhombus (parallelogram)	$A = bh$	The base (b) is the side to which the altitude or height (h) is drawn to perpendicular.
Rectangle (parallelogram)	$A = bh$ or $A = lw$	The base (b) is the side to which the altitude or height (h) is drawn to perpendicular, or use length (l) × width (w).
Square (rectangle)	$A = bh$ or $A = s^2$	The base (b) is the side to which the altitude or height (h) is drawn to perpendicular, or square the side.
Trapezoid	$A = \frac{1}{2}h(b_1 + b_2)$	In a trapezoid the bases are not the same length. One base is b_1 and the other is b_2. The perpendicular distance between the bases is the altitude or height (h).
Circle	$A = \pi r^2$	The radius of the circle is r. For π, use 3.14 or $\frac{22}{7}$.

Meeting the Challenge...Mathematics

5. The **slant height** of a cone or pyramid is the hypotenuse of a right triangle whose legs are the **altitude** of the cone or pyramid and the segment representing the distance from the center (where the altitude meets the base) to the edge of the figure.

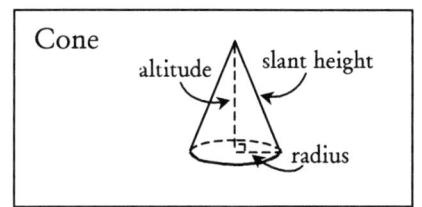

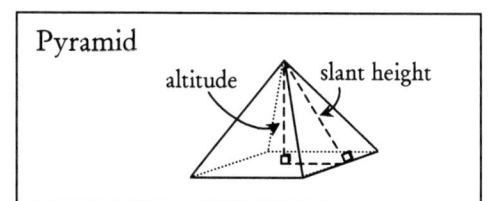

Cone — altitude, slant height, radius

Pyramid — altitude, slant height

6. Use the following formulas to find the surface areas and the volumes of these three-dimensional objects. These formulas are also included on the formula sheet at the back of this study guide. Refer to this formula sheet just as you would during the actual test. Keep in mind that in working with solids, the **surface area** is composed of the lateral area plus the area of the base.
   - The **Lateral Area** (LA) is the sum of the areas of the lateral faces.
   - The **area of the base figure** would be the area of the two-dimensional figure that serves as the base of the prism.

Three-dimensional Objects	Volume Formula	Surface Area Formula	Explanation
Right Prism (with any base)	V = area of base × h	SA = LA + (2 × area of base)	In a right prism, the lateral faces are perpendicular to the two-dimensional figures that are the bases.
Triangular Prism	V = area of △ base × h	SA = LA + (2 × area of △ base)	The bases of the prism are triangles. The lateral faces are rectangles with varying dimensions depending on the sides of the triangular base.
Rectangular (solid) Prism	V = area of ▭ base × h	SA = LA + (2 × area of ▭ base)	The base of the prism is a rectangle. The lateral faces are also rectangles with varying dimensions depending on the sides of the rectangular base.
Pyramid	$V = \frac{1}{3} \times$ area of base $\times h$	SA = LA + area of base	The base of a pyramid can be any regular polygon. The altitude of the pyramid meets the base at the center of the polygon. The lateral faces are congruent isosceles triangles with the slant height of the pyramid being the altitude of the isosceles triangle.
Cylinder	V = area of ○ base × h	SA = LA + (2 × area of ○ base)	The base is a circle. LA = Circumference × h
Cone	$V = \frac{1}{3} \times$ area of ○ base $\times h$	n/a	The altitude of the cone is h.
Sphere	$V = \frac{4}{3} \times \pi \times r^3$	n/a	The radius of the sphere is r.

© 2003 Orange Frazer Press. All Rights Reserved.

BE ABLE TO PERFORM THESE OPERATIONS:

Choosing appropriate units of measure

1. Measurements of length are one-dimensional with the unit of measure in inches, feet, meters, miles, etc.
2. Area measure is two-dimensional involving a multiplication of two dimensions. The unit of measure is in square units, such as square feet, square meters, square miles, etc.
3. Volume measure is three-dimensional involving multiplication using three dimensions. The unit of measure is in cubic units such as cubic inches, cubic centimeters, cubic feet, etc.
   Example: The answer to a problem is 515 square yards. Has length, perimeter, area, surface area or volume been calculated?
   Answer: area or surface area

Finding lengths of segments in figures

1. **Find the distance between two points on a coordinate graph.**
   Use either the distance formula or the Pythagorean theorem (#2 below).
   Example: Find the shortest distance between the two points on the graph.
   Solution: Distance $= \sqrt{(3-1)^2 + (2-(-4))^2} = \sqrt{(2)^2 + (6)^2} = \sqrt{40}$
   Answer: Distance = 6.32 units

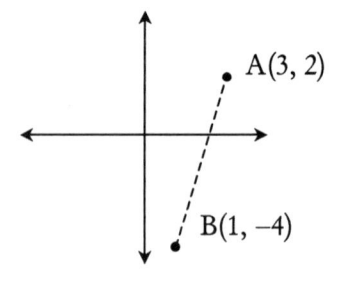

2. **Find the sides of a right triangle using the Pythagorean theorem.**
   Example 1:
   Solution: $c^2 = 7^2 + 3^2 = 49 + 9 = 58$ 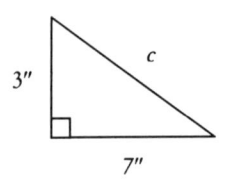
   Answer: $c = \sqrt{58} = 7.62$ inches

   Example 2: 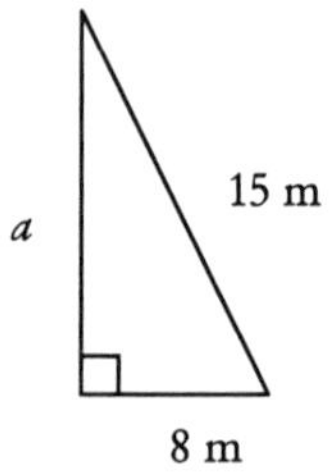
   Solution: $15^2 = 8^2 + a^2$
   $225 = 64 + a^2$
   $a^2 = 225 - 64 = 161$
   Answer: $a = \sqrt{161} = 12.69$ meters

3. **Solve problems involving the circumference of a circle.**
   Example 1: Find the circumference of a circle with a radius of 16 feet.
   Solution: Use the radius to find the diameter.
   $d = 16 \times 2 = 32$ feet
   Answer: $C = \pi \times 32 = 100.48$ feet

   Example 2: If the circumference of a circle is 25 meters, find the diameter.
   Solution: Use the circumference formula.
   $25 = \pi \times d$
   Answer: $d = 25 \div \pi = 7.96$ meters

Meeting the Challenge...Mathematics

Finding areas of two-dimensional figures

Use the area formulas that you have memorized or refer to the formula sheet.

1. **Triangle**
   Example: Find the area.
   Solution: $A = \frac{1}{2} \times 8 \times 6$
   Answer: $A = 24$ square feet

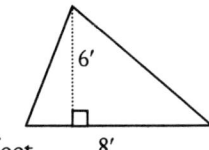

2. **Parallelogram or rhombus**
   Example: Find the area.
   Solution: $A = 16 \times 7$
   Answer: $A = 112$ square yards

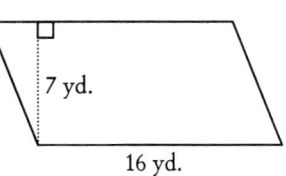

3. **Rectangle or square**
   Example: Find the area.
   Solution: $A = 4 \times 7$
   Answer: $A = 28$ square miles

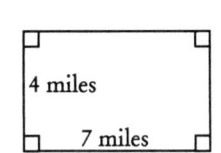

4. **Trapezoid**
   Example: Find the area.
   Solution: $A = \frac{1}{2} \times 10 \times (22 + 18)$
   Answer: 200 square centimeters

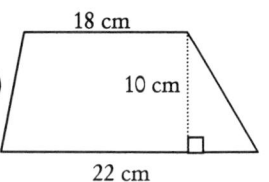

5. **Circle**
   Example: Find the area.
   Solution: $A = \pi \times 9^2$
   Answer: $A = \pi \times 81 = 254.34$ square inches

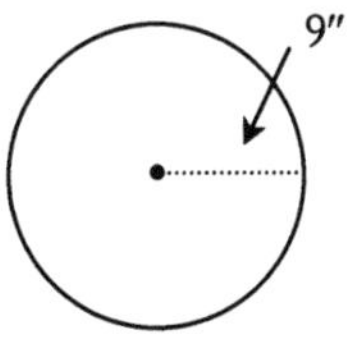

6. **Find combined or cut out areas.**
   Example: Find the total area of this figure.
   Solution: 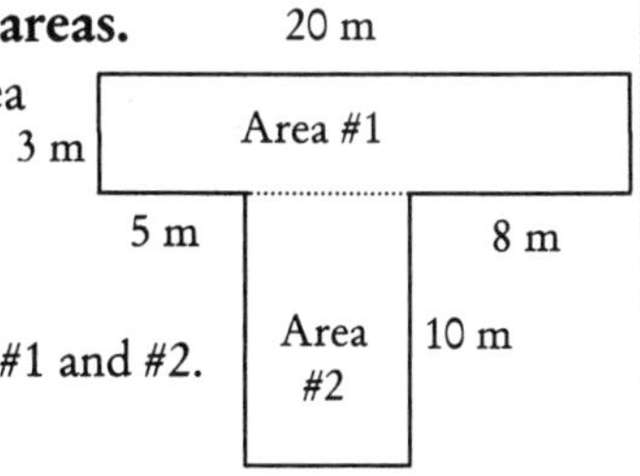
   Subdivide the figure into areas #1 and #2.
   Area #1 = $20 \times 3 = 60$
   Area #2 = $10 \times 7 = 70$
   Answer: The combined area is 130 square meters.

Finding missing parts of two-dimensional figures

If the area of a figure is given, use the area formula for that figure to find the missing part.
Example: The area of a Δ is 38 square inches and the base is 10 inches. Find the altitude.
Solution: Substitute the given values into the formula for the area of a Δ ($A = \frac{1}{2} bh$) and solve for the altitude (h).
Answer: $38 = \frac{1}{2} \times 10 \times h$ $\qquad$ $38 = 5h$ $\qquad$ $h = 7\frac{3}{5}$ or 7.6 inches

Finding the volume of three-dimensional objects

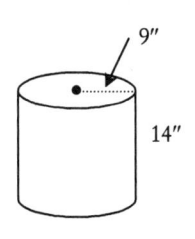

Use the volume formulas that you have memorized or refer to the formula sheet.
Example: Find the volume of the figure.
Solution: Identify the figure as a cylinder. Substitute the given values into the volume formula. V = area of the base(circle) × height
Answer: $V = \pi \times 9^2 \times 14$ $\qquad$ $V = 3.14 \times 81 \times 14$ $\qquad$ $V = 3560.76$ cubic inches

Finding missing parts of three-dimensional objects

If the volume of a figure is given, use the volume formula for that object to find the missing part.
Example: The volume of a cone is 90 cubic centimeters and the height (altitude) is 10 centimeters. Find the radius.
Solution: This is a two part problem.
  a) First, find the area of the base. Substitute the given values into the volume formula. $V = \frac{1}{3} \times B \times h$
     Answer: $90 = \frac{1}{3} \times B \times 10$ $90 = \frac{10}{3} \times B$ $90 \times \frac{3}{10} = B$ $B = 27$ square centimeters
  b) Second, find the radius of the circular base. The area of the base is 27 square centimeters, so use the formula for the area of a circle to solve for the radius. $A = \pi \cdot r^2$
     Answer: $27 = 3.14 \times r^2$ $\frac{27}{3.14} = r^2$ $8.6 = r^2$ $r = \sqrt{8.6} = 2.93$ centimeters

Finding the surface area of three-dimensional objects

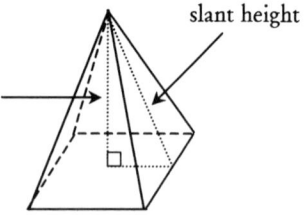

1. **Find the slant height and then find the surface area of a pyramid.**
   Example: The altitude of the pyramid is 7 inches (one leg of the right triangle). Half the side of the square base is 3 inches and serves as the other leg of the right triangle with the slant height as the hypotenuse.
     a) To find the slant height, use the Pythagorean theorem to find the hypotenuse.
        Answer: (slant height)$^2 = 7^2 + 3^2 = 49 + 9 = 58$ Slant height $= \sqrt{58} = 7.62$ inches
     b) To find the surface area, use the formula LA + area of the base.
        To solve for LA, first solve for the area of one triangular lateral face. $A = \frac{1}{2} \times 6 \times 7.62 = 22.86$ sq. in.
        LA $= 4 \times 22.86 = 91.44$ square inches
        Area of the base $= 36$ square inches
        Answer: Surface area $= 91.44 + 36 = 127.44$ square inches

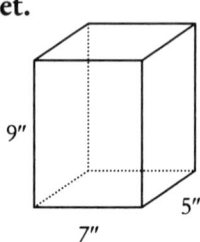

2. **Use the surface area formulas that you have memorized or refer to the formula sheet.**
   Example: Find the surface area of a rectangular solid (prism).
   Solution: Six lateral and base areas make up the surface area of a rectangular solid.
        There are 2 lateral faces, right and left, with dimensions 9×5.
        There are 2 lateral faces, front and back, with dimensions 9×7.
        There are 2 bases, top and bottom, with dimensions 7×5.
   Answer: Total surface area $= 90 + 126 + 70 = 286$ square inches

BE ABLE TO PERFORM THESE CALCULATOR OPERATIONS:

The Pythagorean theorem and the distance formula will require the use of $\boxed{x^2}$ and $\boxed{\sqrt{x}}$.

Meeting the Challenge...Mathematics

GUIDED PRACTICE

A) **In the following problems, are you solving for length, area, surface area or volume?**

1) concrete needed to pour a foundation _____

2) fabric needed to cover a table _____

3) distance traveled from Euclid to Dayton _____

4) wrapping paper to cover a box _____

5) paint needed for a room _____

6) space in a refrigerator _____

B) **Use the Pythagorean theorem to answer these questions.**

7) Is this a right triangle?
   Yes   No

   Use the Pythagorean theorem to verify your answer below.

   5, 13, 12

8) Is a triangle with 6", 8", and 12" sides a right triangle? _____

   Use the Pythagorean theorem to verify your answer.

9) Solve for c.

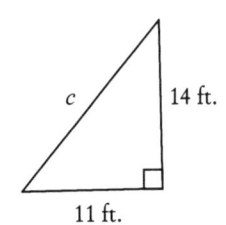

   c, 14 ft., 11 ft.

   $c =$ _____

10) Solve for b.

    15.2 yds., 22.9 yds., b

    $b =$ _____

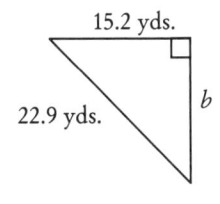

C) **Find the perimeter and the area of the following figures.**

11) Parallelogram

    12", 15", 18"

    a) perimeter = _____   b) area = _____

12) Trapezoid

    10", 25", 15", 17", 38"

    a) perimeter = _____   b) area = _____

© 2003 Orange Frazer Press. All Rights Reserved.

13) Rectangle

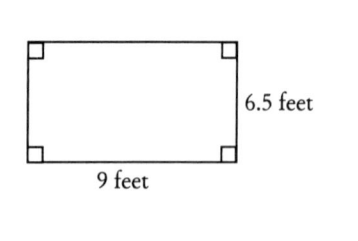

6.5 feet
9 feet

a) perimeter = _____ b) area = _____

14) Triangle

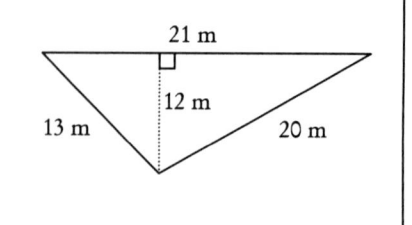

21 m
12 m
13 m   20 m

a) perimeter = _____ b) area = _____

D) Find the circumference and the area in the following figures.

15) Circle

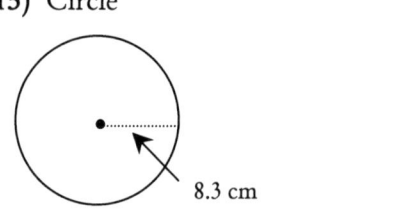

8.3 cm

a) area = _____
b) area of a semicircle = _____
c) circumference = _____

16) Circle

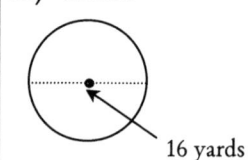

16 yards

a) area = _____
b) area of a quarter circle = _____
c) circumference = _____

E) Solve the following problems involving surface area and volume.

17) A rectangular prism is shown. Find the area of the front/back, right/left sides, and top/bottom. Then find the surface area and volume.

10 cm
8 cm
3 cm

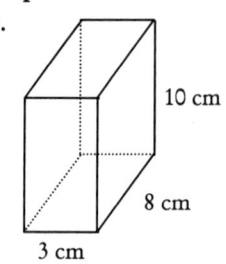

a) area of front or back = _____
b) area of right or left side = _____
c) area of top or bottom = _____
d) total surface area = _____
e) volume = _____

18) Find the area of the top/bottom and the lateral area. Then find the surface area and volume.

7 ft.
21 ft.

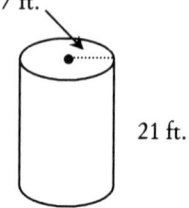

a) area of top or bottom = _____
b) circumference = _____
c) lateral area = _____
d) total surface area = _____
e) volume = _____

44

Meeting the Challenge...Mathematics

19) This pyramid has a square base. The length of the side of the square is 12 meters. The length of the altitude is 8 meters. Find the surface area and volume.

To find the surface area, solve the following:

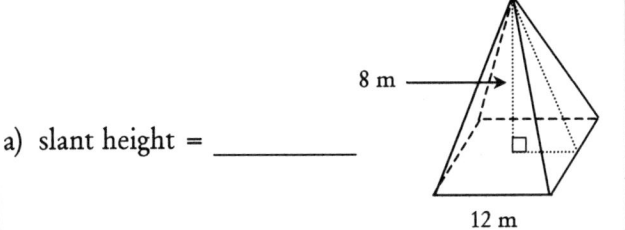

a) slant height = _____

b) area of a lateral face = _____

c) total lateral area = _____

d) area of the base = _____

e) total surface area = _____

volume = _____ = _____

20) This cone has an altitude of 12 inches and the base has a radius of 5 inches. Find the volume.

To find the volume, solve the following:

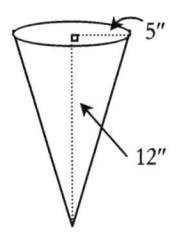

a) area of the base = _____

b) volume = _____

21) Find the total area by subdividing this figure and using combinations of areas. Show your work on the figure.

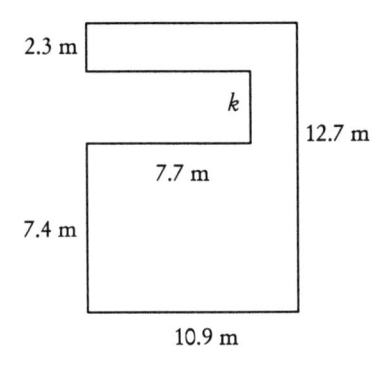

2.3 m
k
12.7 m
7.7 m
7.4 m
10.9 m

k = _____   area = _____

22) Find the area of this figure by using combinations of areas. Show your work on the figure.

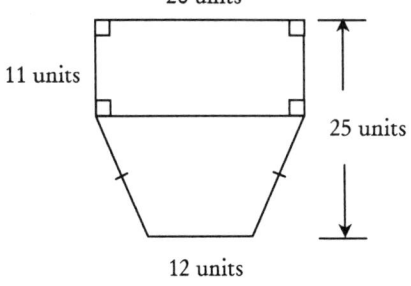

20 units
11 units
25 units
12 units

area = _____

INDEPENDENT PRACTICE

A) Solve the following problems using the Pythagorean theorem. If necessary, round answers to two decimal places.

1) Is this a right triangle?   Yes   No

   Use the Pythagorean theorem to verify your answer.

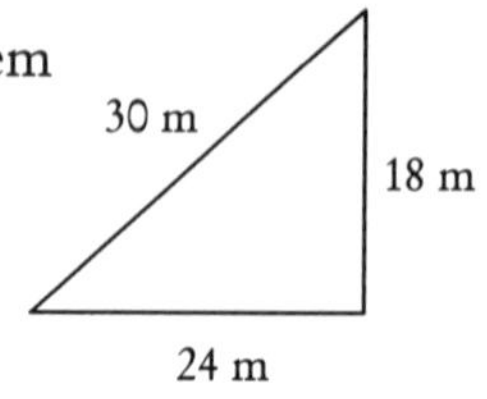

   30 m, 18 m, 24 m

2) Find the missing length.

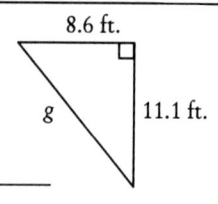

   8.6 ft., g, 11.1 ft.

   a) The missing length is which side of the right triangle? _____
   b) Show the Pythagorean theorem substitution. _____
   c) Answer: g = _____

3) Find the missing length.

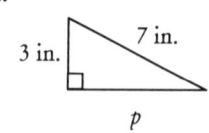

   3 in., 7 in., p

   a) The missing length is which side of the right triangle? _____
   b) Show the Pythagorean theorem substitution. _____
   c) Answer: p = _____

4) Find the missing length.

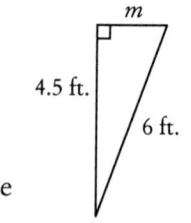

   m, 4.5 ft., 6 ft.

   a) The missing length is which side of the right triangle? _____
   b) Show the Pythagorean theorem substitution. _____
   c) Answer: m = _____

B) Use the distance formula to find the length of the segment that would connect the given points. Show your formula work. If necessary, round answers to two decimal places.

Points: R(−4, 2), S(2, 5), T(−2, −2), V(1, −5)

5) RT = _____ units

6) TV = _____ units

7) from S to the origin = _____ units

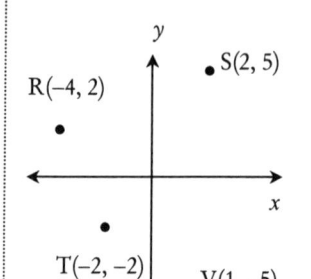

8) from the origin to R = _____ units

9) VS = _____ units

10) RS = _____ units

11) perimeter of quadrilateral RTVS = _____ units

46

Meeting the Challenge...Mathematics

C) Find the missing lengths in these circles. If necessary, round answers to two decimal places.

12) 8.4 ft.

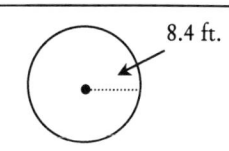

a) radius = _____

b) diameter = _____

c) circumference = _____

13) 16 in.

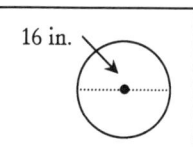

a) radius = _____

b) diameter = _____

c) circumference = _____

14) The circumference of a circle is 12 m.

a) radius = _____

b) diameter = _____

D) Find the area of these two-dimensional figures. If necessary, round answers to two decimal places.

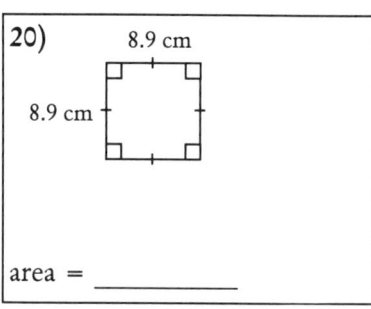

15) 35 m, 40 m, 60 m

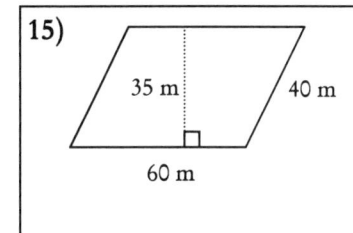

area = _____

16) 15 cm

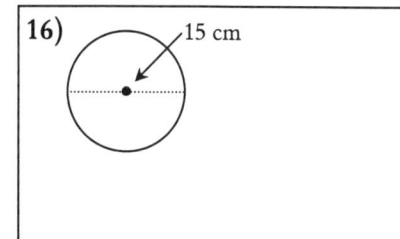

area = _____

17) 7 ft., 9 ft., 20 ft.

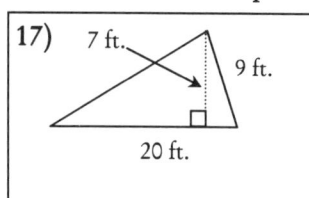

area = _____

18) 7.4 inches, 2.9 inches

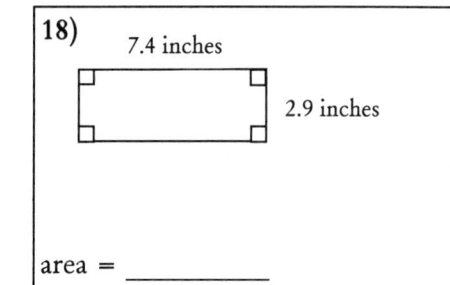

area = _____

19) 18 ft., 12 ft., 30 ft.

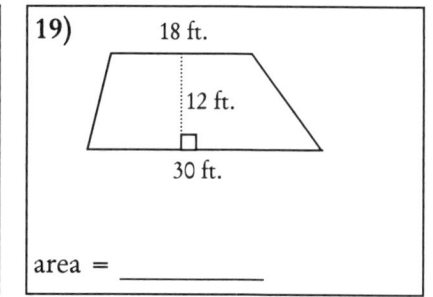

area = _____

20) 8.9 cm, 8.9 cm

area = _____

21) Find the shaded area.
30 in., 16 in., 6 in., 20 in.

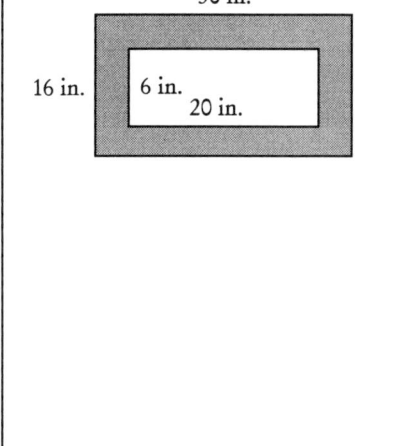

shaded area = _____

22) Find the shaded area.
14 ft., 6 ft.

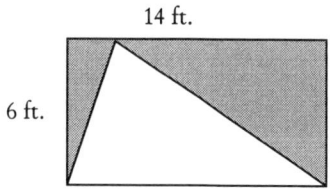

shaded area = _____

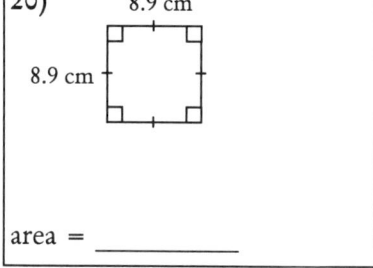

23) Find the combined area.
12 yds., 8 yds.

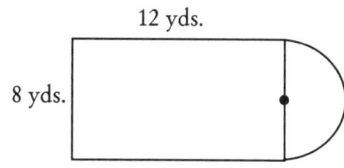

combined area = _____

24) Find the total area.

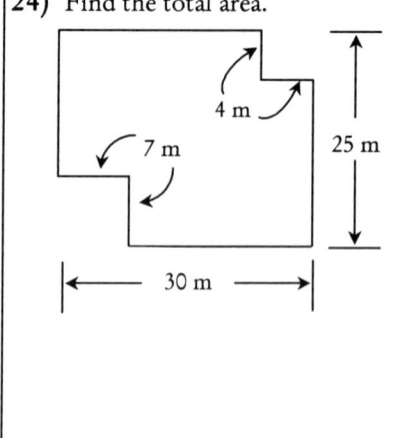

7 m, 4 m, 25 m, 30 m

total area = _____

25) Find the shaded area.

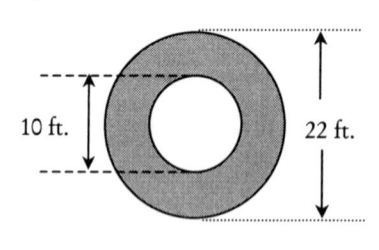

10 ft. 22 ft.

shaded area = _____

26) This square has 17 cm sides. Find the shaded area.

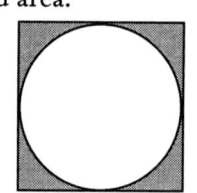

shaded area = _____

E) **Find the surface area and/or the volume.** If necessary, round answers to two decimal places.

27) Show formulas, substitutions and solutions.

Volume

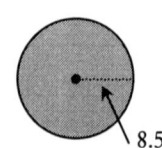

8.5 yds.

volume = _____

28) Show formulas, substitutions and solutions.

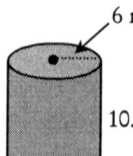

Surface Area   6 m   Volume

10.5 m

surface area = _____   volume = _____

29) Show formulas, substitutions and solutions.

Surface Area   Volume

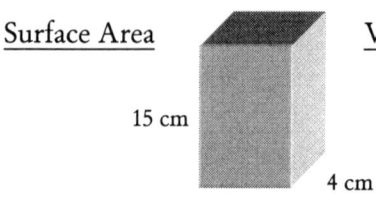

15 cm, 4 cm, 6 cm

surface area = _____   volume = _____

30) Show formulas, substitutions and solutions.

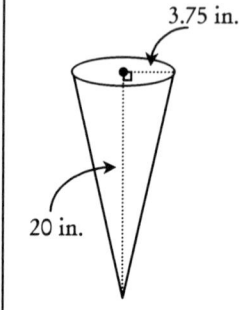

3.75 in.

Volume

20 in.

volume = _____

Meeting the Challenge...Mathematics

F) Solve the following application problems. If necessary, round answers to two decimal places.

31) The high school cross country practice course runs 2 km east, 3 km north, 1 km west, and 5 km north. The runners then return to the start/finish line on a direct route as the crow flies.

　a) Draw a sketch of the course. Label the start/finish line, the turn around point and distances.

　b) Find the 'direct route' distance from the turn around point back to the finish line.

　　'direct route' distance = _____

　c) Calculate the total running distance of the course.

　　total running distance = _____

32) How many cubic inches of dirt are needed to fill this planter? The end pieces of this planter are semicircles.　30"　6"

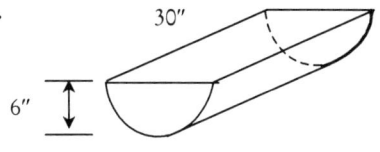

　dirt needed = _____

33) A snow cone cup is 4 inches deep and 2 inches across the top. A single scoop of flavored ice 2 inches in diameter is placed in the cone. If the ice melts, will the cone hold all the liquid? _____

Justify your answer by showing formulas, substitutions and solutions.

34) Find the area of this piece of land. Subdivide the figure on the sketch and find the area of each part. Show all your formula work, substitutions and solutions.

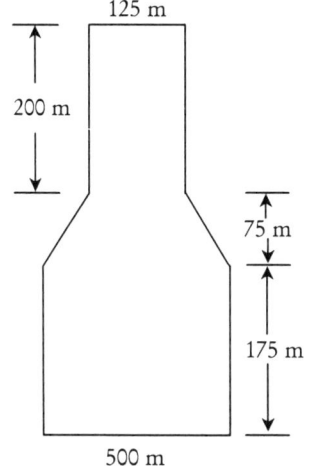

125 m　200 m　75 m　175 m　500 m

total area of the land = _____

TEST PRACTICE

MULTIPLE-CHOICE QUESTIONS:

Choose the correct answer. If necessary, you may round answers to one or two decimal places as dictated by the answer choices.

1) A 25-foot ladder is leaning against a wall. The foot of the ladder is 5 feet from the base of the wall. How far up the wall will the ladder reach?

   a. 20 feet   b. $24\frac{1}{2}$ feet   c. 22.5 feet   d. $25\frac{1}{2}$ feet

2) What is the area of a circle with a diameter of 14 cm?

   a. 43.96 square cm   b. 28.26 square cm   c. 153.86 square cm   d. 104.76 square cm

3) A cone and a cylinder have the same volume. The cone has a radius of 7.8 inches and a height of 5.2 inches. The cylinder has a radius of 3.9 inches. The height of the cylinder is approximately

   a. 7 inches   b. 12 inches   c. 10 inches   d. 15 inches

4) Find the length of the segment with endpoints K(–4, –3) and M(3, –1).

   a. $\sqrt{65}$ units   b. $\sqrt{53}$ units   c. $\sqrt{17}$ units   d. $\sqrt{33}$ units

5) How much wrapping paper will it take to cover a rectangular box with dimensions of 9 in. by 13 in. by 5 in.? Do not consider any wasted paper in your calculations.

   a. 415 square inches   b. 454 square inches   c. 227 square inches   d. 585 square inches

SHORT ANSWER QUESTIONS:

6) When the reserve football field is open, students are permitted to take a shortcut diagonally across the field.

   a) Show your work to set up and calculate the length of the shortcut across the football field.

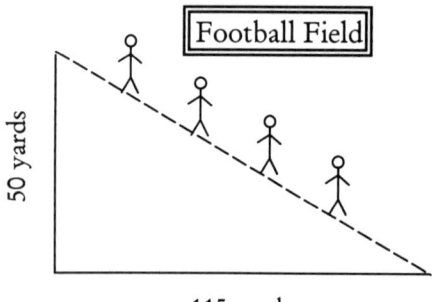

   [Football Field diagram: right triangle with legs 50 yards and 115 yards, dashed diagonal shortcut]

   b) Find the walking distance the shortcut will save.

Meeting the Challenge...Mathematics

7) Consider △ABC, △ABD and △ABE.

What can you say about the areas of these three triangles? _____

Why? _____
_____

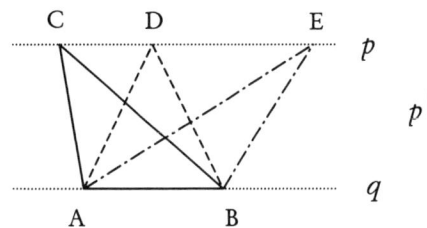

8) Gary needs to rent a storage unit. He has checked out two units that rent for $50 a month but he wants to make sure he rents the unit with the larger storage capacity. The dimensions of the unit at Ace Storage are 15 ft. × 10 ft. × 12 ft. The dimensions of the unit at Mike's Storage are 20 ft. × 8 ft. × 11 ft. Determine which unit has the most space. Justify your answer.

Which company's unit has more storage space? _____

EXTENDED RESPONSE QUESTIONS:

9) Coffee cans with plastic lids are going to be used as little drums at the kindergarten class party. The cans will be wrapped with patriotic fabric for the occasion.

a) Explain how to determine the amount of fabric needed to wrap one can. The top and bottom will not be wrapped.
_____
_____

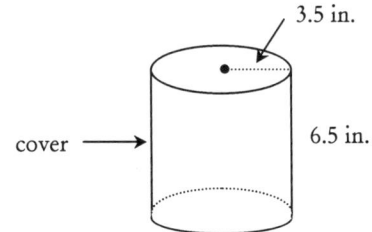

b) Find the dimensions of the fabric needed to cover one coffee can.

dimensions of fabric: _____

c) Calculate the amount of fabric needed to cover one coffee can.

fabric needed in square inches: _____

10) Rhonda has a square patio. She would like to double the area of the patio and decides to do it by doubling the length and the width. Will her idea work? _____

Justify your answer with examples, sketches, formulas and explanation.

CLOSURE

1) The size of the circular pool cover would be the diameter of 20 ft. plus an additional foot to hang over the rim. This adds one foot onto each end of the diameter.
   Answer: a circular cover with a diameter of 22 feet

2) To find the cost of the pool cover, first calculate the area of the circular pool cover.
   Area = $\pi \times r^2$ = 3.14×11^2 = 3.14×121 = 379.94 square feet
   Then multiply the area of the pool cover by the cost per square foot. **Cost:** 379.94 × $0.89 = **$338.15**

3) The pool is actually a cylinder with a diameter of 20 ft. and a depth of 4 ft. Use the volume formula to find the number of cubic feet of water. V = area of the circular base × depth
   V = $\pi \times r^2 \times$ depth = $3.14 \times 10^2 \times 4$ The pool will hold 1256 cubic feet of water.

4) To find the cost of the water to fill the pool: 1256 cubic feet × $0.18 = $226.08

5) Refer to the sketch. The circular privacy fence has a diameter of 26 feet. (20 + 3 on each end). The circumference of this circle will tell us the amount of fencing needed.
   C = $\pi \times d$ = 3.14×26 = 81.64 feet

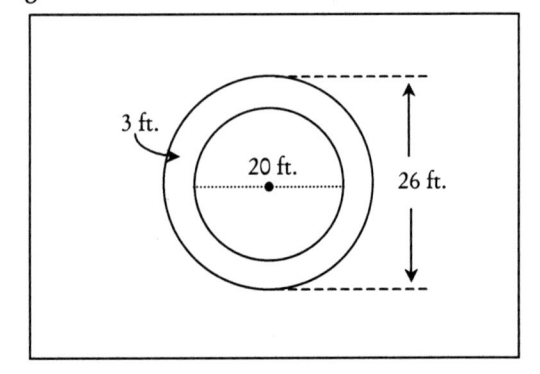

Meeting the Challenge...Mathematics

MATHEMATICAL PROCESSES
Visualizing to Solve Problems

PURPOSE: In this lesson you will use a variety of appropriate mathematical representations to organize, record and communicate mathematical ideas.

A) During the fall harvest, the Heck Grain Storage Company ran out of storage space and started dumping corn onto the pavement. The corn dropped vertically from a chute 50 feet above the ground. After several days, a huge cone-shaped pile of corn had formed. The cone was 45 feet high. The circumference of the circular base of the pile was 180 feet.

Directions and questions:
Show formulas and formula substitutions, sketches and dimensions. Explain your logic.

1) Make a sketch of the corn pile and show the necessary dimensions.

2) Find the volume of the corn pile in cubic feet.

_____ cubic feet

3) Knowing that a bushel is approximately 1.24 cubic feet, calculate the bushels of corn in the pile.

_____ bushels

4) The insurance company needs to know the value of the corn pile. Go to the Internet or check the local television news or newspaper to find the current price of a bushel of corn.

value of the corn pile _____

corn pile

B) A rectangular pan has dimensions of 6 in. × 8 in. × 10 in. A cylindrical glass has a diameter of 7 in. and is 10 in. high. Will the pan hold a full glass of water? _____

1) Before answering this question, sketch each object and show the dimensions.

2) Discuss what information needs to be determined.

_____

_____

3) Solve for this needed information.

sketch

© 2003 Orange Frazer Press. All Rights Reserved.

C) Two airplanes leave the Columbus International Airport at the same time, one traveling due north at a ground speed of 400 m.p.h. and the other traveling due east at a ground speed of 325 m.p.h.

1) Trace the paths of the two planes on the graph. Show how far they traveled after three hours and where they would be in relationship to each other.

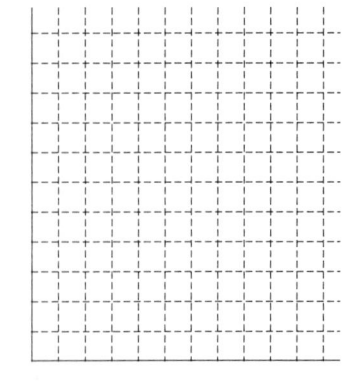

2) Explain how to find how far the planes are from each other after 1.5 hours. Represent the distance on the graph and solve for the distance.

3) Explain how to find how far the planes are from each other after 3 hours. Represent the distance on the graph and solve for the distance.

D) Using the blueprint of a house, Bruce needs to find enough information to determine the dimensions of the roof.

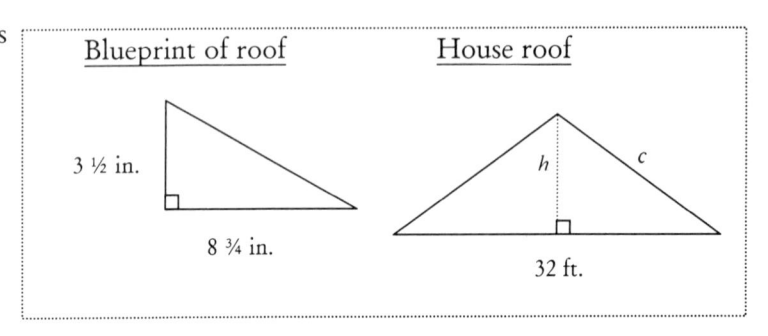

Blueprint of roof: $3 \frac{1}{2}$ in., $8 \frac{3}{4}$ in.

House roof: h, c, 32 ft.

1) Construct a proportion to find the value for h.

2) Solve for h.

3) The contractor is ready to order materials to roof the house. Explain how to find side c by including a formula with your explanation.

4) Solve for side c.

5) The length of the house is 40 feet. Find the number of square feet that need to be roofed.

6) The overall cost to roof a house is approximately $4 per square foot. Find the approximate cost to roof this house.

Meeting the Challenge...Mathematics

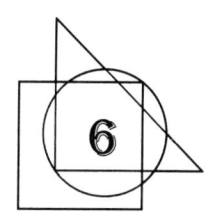

ANGLES AND LINES

Apply angle relationships to situations that involve intersecting, perpendicular and parallel lines.

FOCUS: Our world is filled with many examples of angles and lines.

As you go through your day, look around and list examples of various combinations of lines and angles that make up your world.

PURPOSE: When you have completed this lesson, you will be able to solve problems involving angles and lines, and their relationships.

WHAT YOU NEED TO KNOW

MATHEMATICS AND PROBLEM SOLVING VOCABULARY:

1. Acute angle
2. Adjacent angles
3. Alternate exterior angles
4. Alternate interior angles
5. Angle bisector
6. Complementary angles
7. Corresponding angles
8. Intersecting lines
9. Midpoint
10. Obtuse angle
11. Parallel lines
12. Perpendicular lines
13. Polygon
14. Right angle
15. Straight angle
16. Supplementary angles
17. Transversal
18. Vertex of an angle
19. Vertical angles

MEMORIZE THESE FACTS AND FORMULAS:

1. Identify and use basic geometric terms.

   Point   A dot is used to show a point and is named by using a capital letter.   •B   read: point B

   Segment   A segment is labeled by using the endpoint letters.
   A•——————•B   read: line segment AB or $\overline{AB}$

   Angle   The symbol for an angle is ∠.
   - Angles are named by using a number (∠5), the single letter at the vertex (∠E), or three letters with the middle letter naming the vertex (∠DEF).
   - The measure of an angle is shown as m∠5, m∠E or m∠DEF.
   - Angles are classified by size as follows:

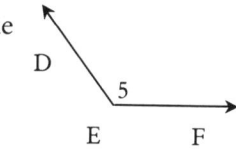

     Acute (between 0° and 90°) 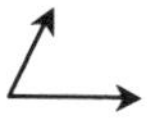
     Right (90°) 
     Obtuse (between 90° and 180°) 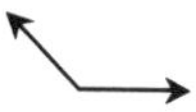
     Straight (180°)

© 2003 Orange Frazer Press. All Rights Reserved.

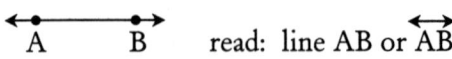

Line A line can be labeled by two points. A B read: line AB or $\overleftrightarrow{AB}$
A line can also be labeled by using a single, lower case letter. m read: line m
The symbol for parallel lines is ∥.
The symbol for perpendicular lines is ⊥.
The symbol for a right angle is ∟.

2. Intersecting lines form four angles. The opposite angles are called **vertical angles** and are equal. m∠1 = m∠3 and m∠2 = m∠4

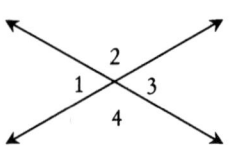

3. **Supplementary angles** are two angles with a sum of 180°.
m∠1 + m∠2 = 180°

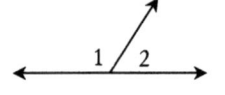

4. **Complementary angles** are two angles with a sum of 90°.
m∠1 + m∠2 = 90°

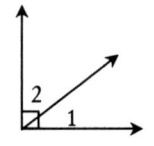

5. The sum of the angle measures in a triangle is 180°. m∠1 + m∠2 + m∠3 = 180°

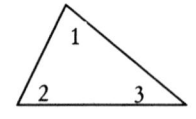

6. When two parallel lines are intersected by a transversal, the following pairs of angles are equal.
   - **Corresponding angles:** m∠1 = m∠5; m∠3 = m∠7; m∠2 = m∠6; m∠4 = m∠8
   - **Alternate interior angles:** m∠3 = m∠6; m∠4 = m∠5
   - **Alternate exterior angles:** m∠1 = m∠8; m∠2 = m∠7

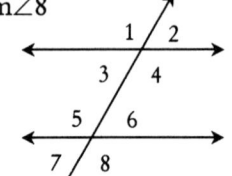

7. **Coordinate geometry** gives position as well as size to segments and angles by using ordered pairs to plot them on the x–y coordinate plane. Use ordered pairs from the endpoints of a segment [P(a, b) and Q(c, d)] to solve for the following:
   a) the length of a segment
   - Use the distance formula. $\sqrt{(a-c)^2 + (b-d)^2}$
   - You can also find the length of a segment by forming a right triangle with that segment being the longest side. This segment is opposite the right angle and is called the **hypotenuse**. Use the **Pythagorean theorem**, $c^2 = a^2 + b^2$, where a and b are the legs of the triangle and c (our segment) is the hypotenuse.

   b) the coordinates of the midpoint of the segment
   Use the midpoint formula. $\left(\dfrac{a+c}{2}, \dfrac{b+d}{2}\right)$

   c) the slope of the segment or line
   Use the slope formula. (See Lesson #12.) $m = \dfrac{b-d}{a-c}$ or $\dfrac{d-b}{c-a}$

Meeting the Challenge...Mathematics

BE ABLE TO PERFORM THESE OPERATIONS:

Identifying angles and relationships

1. Two lines intersect. Identify the following pairs of angles.

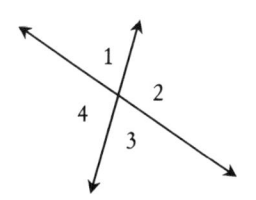

   a) Adjacent angles: ∠1 & ∠4; ∠1 & ∠2; ∠2 & ∠3; ∠3 & ∠4
   b) Vertical angles: ∠1 & ∠3; ∠2 & ∠4
   c) Supplementary angles: ∠1 & ∠4; ∠1 & ∠2; ∠2 & ∠3; ∠3 & ∠4

2. Two lines are intersected by a transversal. Identify the following angles.

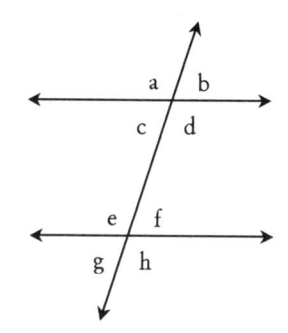

   a) Corresponding angles: ∠a & ∠e; ∠c & ∠g; ∠b & ∠f; ∠d & ∠h
   b) Interior angles: ∠c, ∠d, ∠e, ∠f,
   c) Alternate interior angles by pairs: ∠c & ∠f; ∠d & ∠e
   d) Exterior angles: ∠a, ∠b, ∠g, ∠h
   e) Alternate exterior angles by pairs: ∠a & ∠h; ∠b & ∠g

Finding angle measures

1. Solve for the complement of an angle.
   Example: If m∠1 = 71°, find the complement (x).   Equation: 71° + x = 90°   Answer: x = 19°

2. Solve for the supplement of an angle.
   Example: If m∠1 = 98°, find the supplement (x).   Equation: 98° + x = 180°   Answer: x = 82°

3. Solve for the missing angle measures in a triangle.

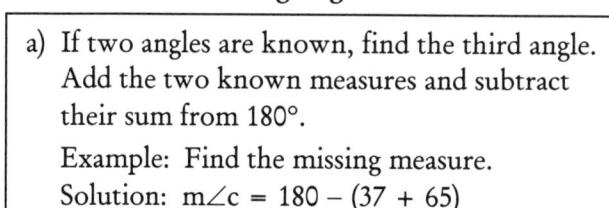

   a) If two angles are known, find the third angle. Add the two known measures and subtract their sum from 180°.
   Example: Find the missing measure.
   Solution: m∠c = 180 − (37 + 65)
   Answer: m∠c = 78°

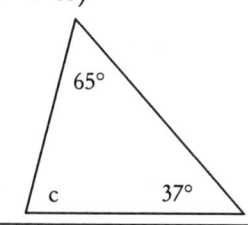

   b) If one angle and an exterior angle not adjacent to the given angle are known, find the measures of the two remaining angles in the triangle. The exterior angle is supplementary to the adjacent angle inside the triangle.
   Example: Find the missing measures.
   Solution: m∠a = 180 − 101 = 79°
   m∠b = 180 − (79 + 48) = 53°
   Answers: m∠a = 79°
   m∠b = 53°

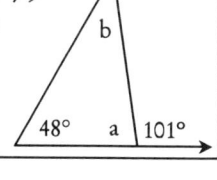

4. Use angle relationships from parallel lines and intersecting lines to find missing angle measures.

   Example: $k \parallel g$ and m∠3 = 123°. Find all missing angle measures.
   Answers: This approach is just one of many different ways to solve this problem.
   a) Since corresponding angles are equal, m∠3 = m∠7 = 123°.
   b) Since alternate interior angles are equal, m∠3 = m∠6 = 123°.
   c) Since vertical angles are equal, m∠3 = m∠2 = 123°.
   d) Since ∠3 is supplementary to ∠1, m∠1 = 57° (180 − 123).
   e) Since alternate exterior angles are equal, m∠1 = m∠8 = 57°.
   f) Since ∠1 and ∠4 are vertical, m∠1 = m∠4 = 57°.
   g) Since ∠8 and ∠5 are vertical, m∠8 = m∠5 = 57°.

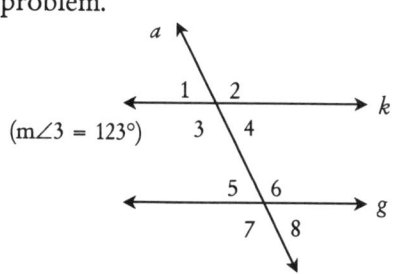

Establishing that lines are parallel

1. **If corresponding angles are equal, then lines are parallel.** If it can be established from the given information that a pair of corresponding angles are equal, then it can be said that the lines are parallel.
2. **If alternate interior angles are equal, then lines are parallel.** If it can be established from the given information that a pair of alternate interior angles are equal, then it can be said that the lines are parallel.
3. **If alternate exterior angles are equal, then lines are parallel.** If it can be established from the given information that a pair of alternate exterior angles are equal, then it can be said that the lines are parallel.

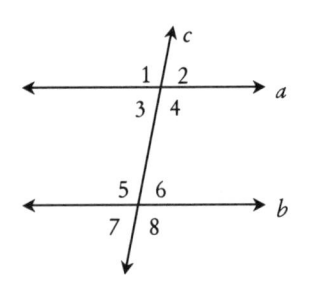

Example: If $\angle 1$ and $\angle 7$ are supplementary, establish that $m\angle 1$ and $m\angle 5$ are equal and corresponding.
Solution: $\angle 1$ and $\angle 7$ are supplementary. $m\angle 1 + m\angle 7 = 180°$
$\angle 5$ and $\angle 7$ are supplementary. $m\angle 5 + m\angle 7 = 180°$
$m\angle 1 + m\angle 7 = m\angle 5 + m\angle 7$ Therefore, $m\angle 1 = m\angle 5$.
Answer: Since $\angle 1$ and $\angle 5$ are corresponding angles, $a \parallel b$.

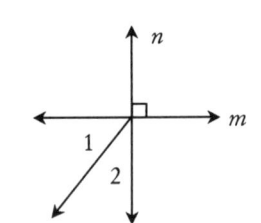

Establishing that lines are perpendicular

1. **If two lines form right angles, the lines are perpendicular.**
   Since line m and line n form right angles, the lines are perpendicular.

2. **If two adjacent angles are complementary, their exterior sides are perpendicular.**
   The given right angle is vertical to the sum of the adjacent angles. $m\angle 1 + m\angle 2 = 90°$
   As a result, we can say that $\angle 1$ and $\angle 2$ are complementary and their exterior sides are perpendicular.

Using coordinate geometry with segments and angles

1. Position a segment on an x-y coordinate graph by using the ordered pairs that name its endpoints.
2. Find the length of a segment by using the distance formula.
3. Find the length of a segment by using the Pythagorean theorem.
4. Find the midpoint of a segment by using the midpoint formula.
   Example: $A(3, 2)$ and $B(1, -2)$ are endpoints of the segment on the graph.
   a) Find the length of segment AB.

   Solution: Use the distance formula. The length of the segment is $\sqrt{(3-1)^2 + (2-(-2))^2}$.
   Answer: $\sqrt{(2)^2 + (4)^2} = \sqrt{20} = 4.47$ units

   Use the Pythagorean theorem ($hyp_c^2 = leg_a^2 + leg_b^2$). Draw the right triangle. $\overline{AB}$ is the hypotenuse.

   Formula: $c^2 = a^2 + b^2$ Substitution: $c^2 = 4^2 + 2^2$ $c^2 = 16 + 4 = 20$ Answer: $c = \sqrt{20} = 4.47$ units

   b) Find the midpoint of $\overline{AB}$. Formula substitution: $\left(\dfrac{3+1}{2}, \dfrac{2+(-2)}{2}\right)$ Answer: midpoint = $(2, 0)$

BE ABLE TO PERFORM THESE CALCULATOR OPERATIONS:

1. Use the square root key. $\boxed{\sqrt{x}}$
2. Use the square key. $\boxed{x^2}$

Meeting the Challenge...Mathematics

GUIDED PRACTICE

A) Tell whether the angles appear to be acute, right, obtuse or straight.

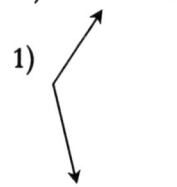

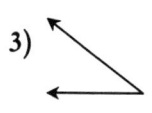

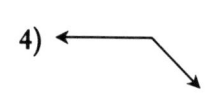

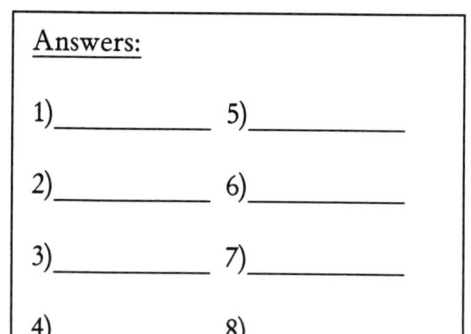

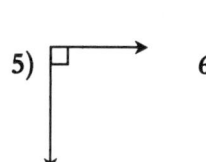

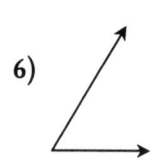

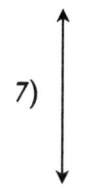

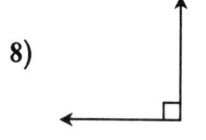

1)   2)   3)   4)

Answers:

1) _____ 5) _____

2) _____ 6) _____

3) _____ 7) _____

4) _____ 8) _____

5)   6)   7)   8)

B) Use each diagram to answer the questions involving angle pairs (adjacent, vertical, supplementary and complementary) and angle sizes (acute, right, obtuse and straight).

9) Fill in the blanks.
   a) ∠1 is adjacent to _____ and _____.
   b) ∠3 is adjacent to _____ and _____.
   c) ∠2 and ∠4 look to be _____ angles.
   d) ∠3 and ∠4 look to be _____ angles.
   e) ∠3 is a (an) _____ angle.
   f) The exterior sides of ∠1 and ∠2 form a (an) _____ angle.
   g) ∠4 is a (an) _____ angle.
   h) If m∠1 = 52°, then the m∠2 = _____°.
   i) If m∠3 = 139°, then the m∠4 = _____°.

10) Fill in the blanks.
    a) One pair of vertical angles is ∠a and _____.
    b) Another pair of vertical angles is ∠d and _____.
    c) ∠a is supplementary to _____ and _____.
    d) ∠d is adjacent to _____ and _____.

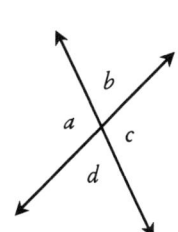

11) From the given angle measures, tell if the angles are complementary, supplementary or neither.

   a) 65° and 105° _____   c) 80° and 100° _____   e) 45° and 45° _____
   b) 60° and 30° _____   d) 40° and 40° _____   f) 107° and 83° _____

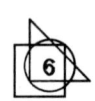

C) **Answer the following questions involving parallel lines intersected by a transversal.**

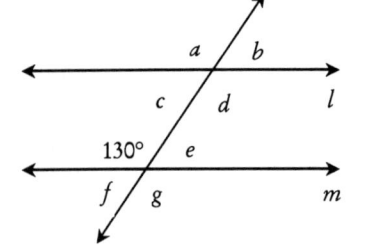

12) It is given that $l \parallel m$.

   a) List the angles with a measure of 130°. _____

   b) The measure of ∠e = _____°.

   c) Tell which angles have the same measure as ∠e. _____

13) It is given that $r \parallel s$.

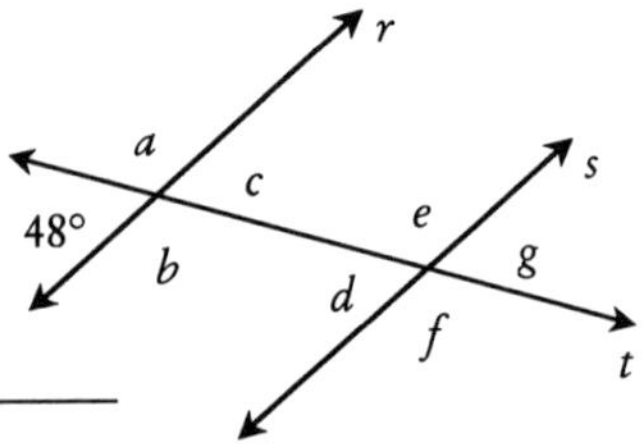

   a) What is the relationship of ∠a and the given 48° angle? _____

   b) What is the measure of ∠a? _____°

   c) List the angles with a measure of 48°. _____

   d) List the angles with a measure the same as ∠a. _____

14) Use only the data given in the diagram.

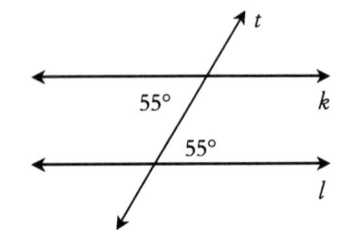

   a) What is the relationship of the two 55° angles? _____

   b) Can it be assumed that $k \parallel l$? _____

   Explain. _____

15) Use only the data given in the diagram.

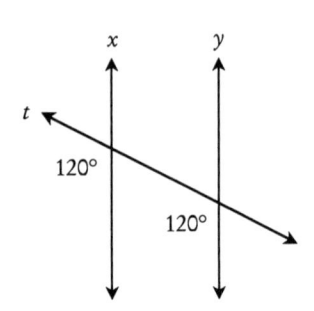

   a) What is the relationship of the two 120° angles? _____

   b) Can it be assumed that $x \parallel y$? _____

D) **Find the angle measures.**

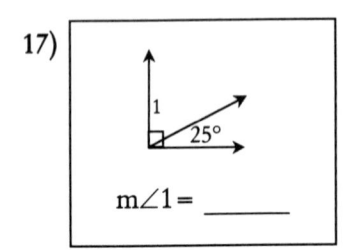

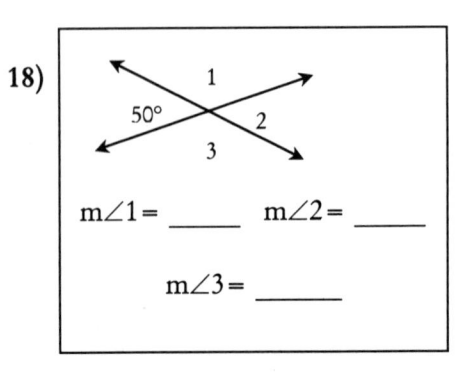

16) 1, 56°; m∠1 = _____

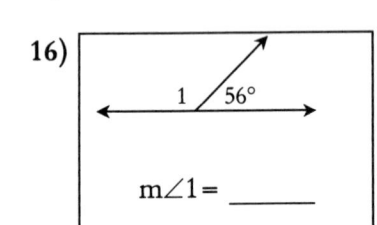

17) 1, 25°; m∠1 = _____

18) 50°, 1, 2, 3; m∠1 = _____ m∠2 = _____ m∠3 = _____

Meeting the Challenge...Mathematics

19)

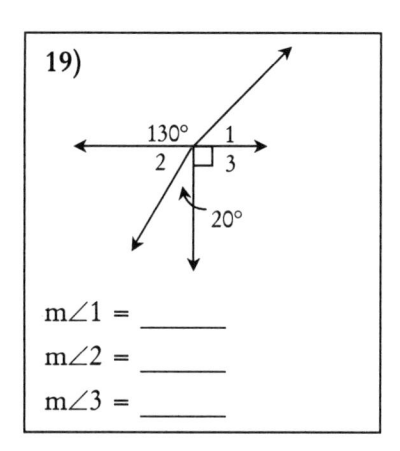

m∠1 = _____
m∠2 = _____
m∠3 = _____

20)

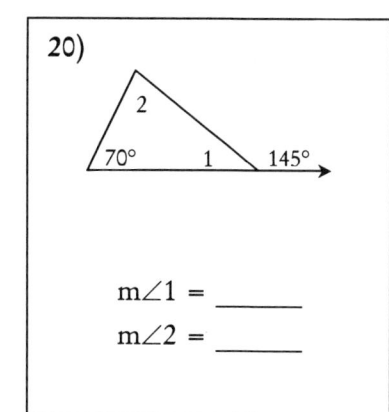

m∠1 = _____
m∠2 = _____

21)

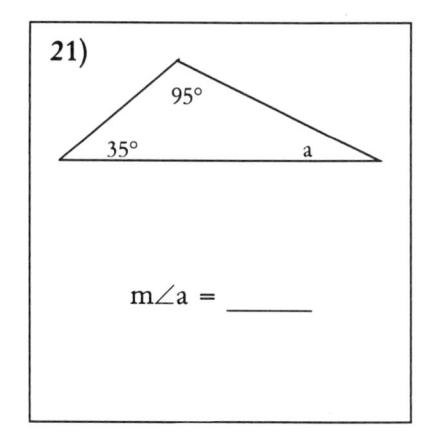

m∠a = _____

E) Use the information on the graph to answer the questions in this section.

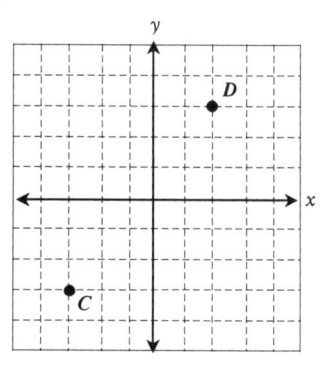

22) Find the length of segment CD using the distance formula.

23) Find the length of segment CD using the Pythagorean theorem.

24) Find the midpoint of $\overline{CD}$.

INDEPENDENT PRACTICE

A) Find the angle measures.

1) $r \parallel t$

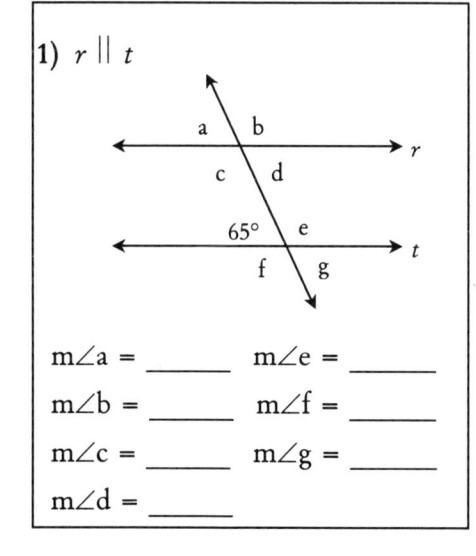

m∠a = _____   m∠e = _____
m∠b = _____   m∠f = _____
m∠c = _____   m∠g = _____
m∠d = _____

2) $h \parallel j$

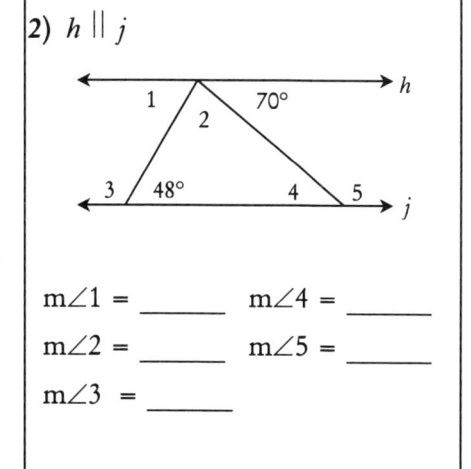

m∠1 = _____   m∠4 = _____
m∠2 = _____   m∠5 = _____
m∠3 = _____

3)

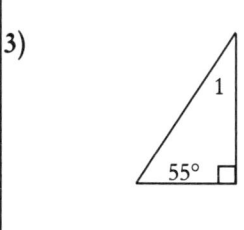

a) m∠1 = _____

b) ∠1 and the 55° angle are _____ angles.

© 2003 Orange Frazer Press. All Rights Reserved.

4) $\overline{AB}$ is an angle bisector.

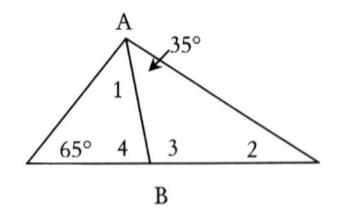

m∠1 = _____
m∠2 = _____
m∠3 = _____
m∠4 = _____

5) $l \parallel \overline{AE}$

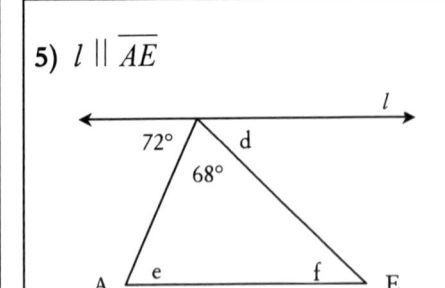

m∠d = _____
m∠e = _____
m∠f = _____

6)

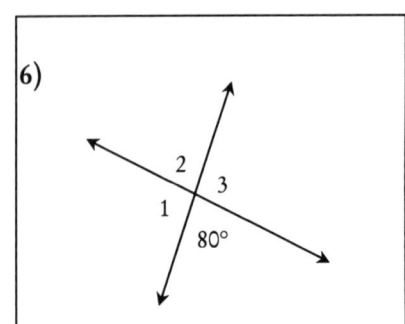

m∠1 = _____
m∠2 = _____
m∠3 = _____

7)

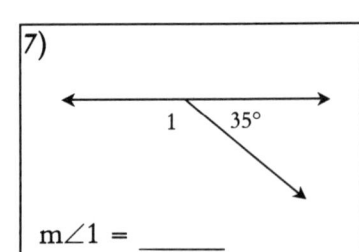

m∠1 = _____

These two angles are _____ and _____.

8)

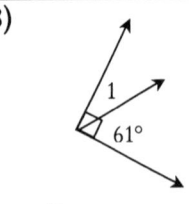

m∠1 = _____

These two angles are _____ and _____.

9) Find the missing values.

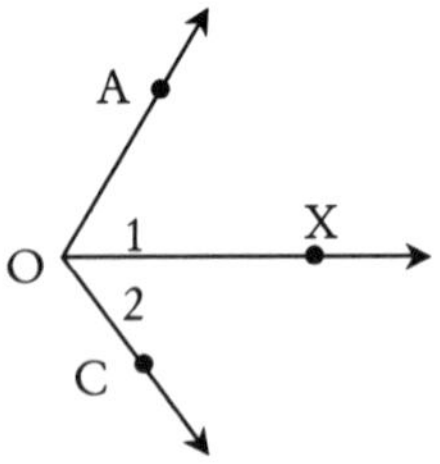

	m∠AOC	m∠1	m∠2
a)		38°	43°
b)	78°	52°	
c)	82°		29°

10) $a \parallel b$

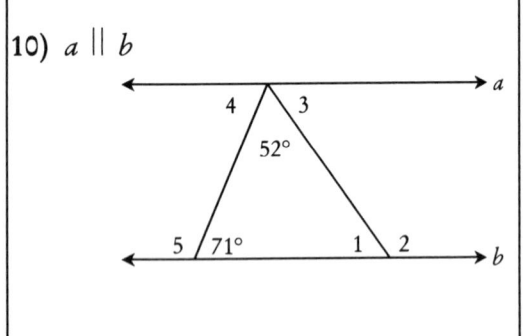

m∠1 = _____   m∠4 = _____
m∠2 = _____   m∠5 = _____
m∠3 = _____

11) $m \parallel n$

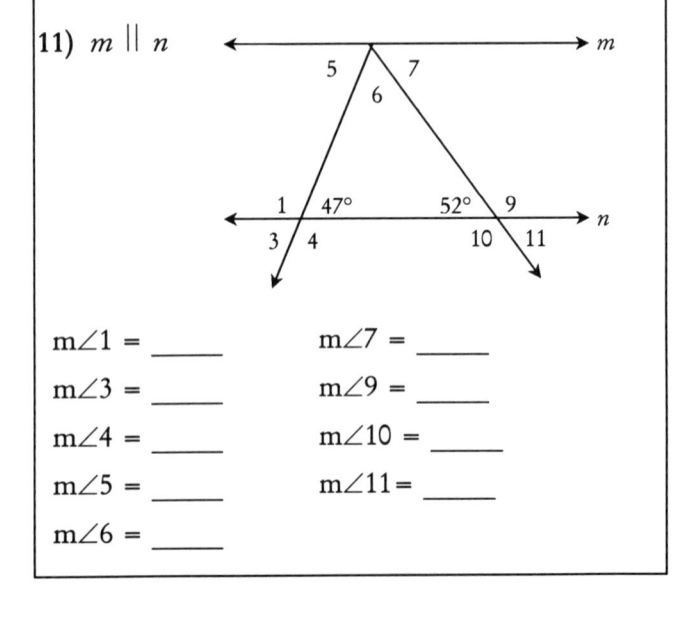

m∠1 = _____   m∠7 = _____
m∠3 = _____   m∠9 = _____
m∠4 = _____   m∠10 = _____
m∠5 = _____   m∠11 = _____
m∠6 = _____

Meeting the Challenge...Mathematics

B) Answer the following questions involving parallel lines intersected by a transversal.

12) Line *k* and line *m* are parallel.

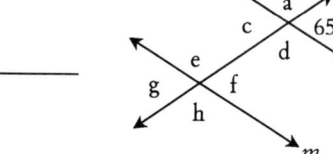

   a) List all the angles with a measure of 65°. _____

   b) The measure of ∠d = _____°.

   c) Tell which angles have the same measure as ∠d. _____

13) It is given that *u* ∥ *v*.

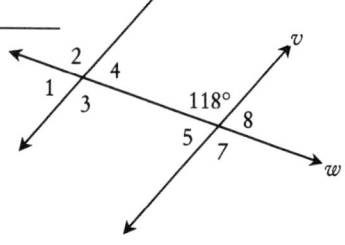

   a) What is the relationship of ∠8 and the 118° angle? _____

   b) What is the measure of ∠8? _____°

   c) List all the angles with a measure of 118°. _____

   d) List all the angles with a measure the same as ∠8. _____

14) Use only the data given in the diagram.

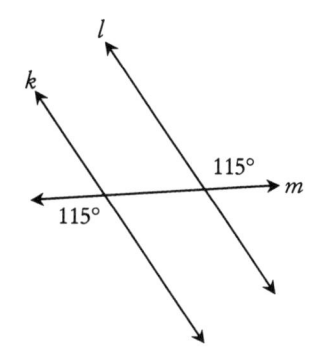

   a) What is the relationship of the two 115° angles? _____

   b) Can it be assumed that *k* ∥ *l*? _____

   Explain. _____

15) Use only the data given in the diagram.

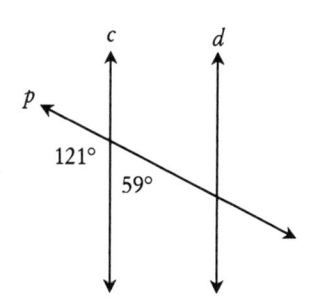

   a) What is the relationship of the two given angles? _____

   b) Can it be assumed that line *c* ∥ *d*? _____

C) Using the given information, can we say without a doubt that *s* ∥ *t*?

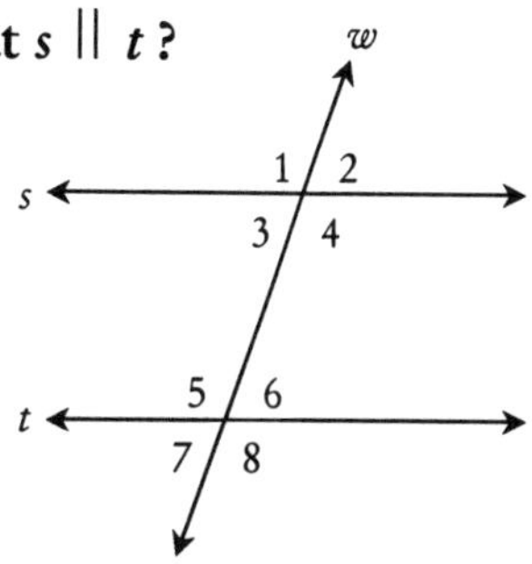

16) m∠2 = m∠3   Yes   No

17) m∠3 = m∠6   Yes   No

18) m∠1 = m∠5   Yes   No

19) m∠5 + m∠6 = 180°   Yes   No

© 2003 Orange Frazer Press. All Rights Reserved.

D) Use the information on the graph to answer the questions in this section.

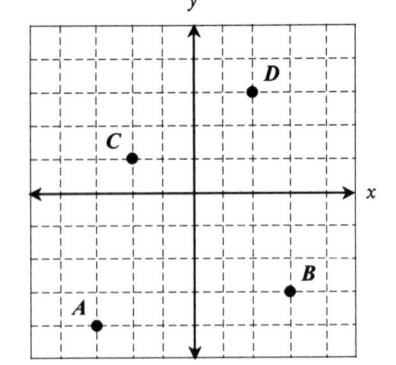

20) Find the length of $\overline{CD}$ using the distance formula.

21) Find the length of $\overline{CD}$ using the Pythagorean theorem.

22) Find the midpoint of $\overline{CD}$.

23) Find the length of $\overline{AC}$ using the distance formula.

24) Find the length of $\overline{BD}$ using the Pythagorean theorem.

25) Find the midpoint of $\overline{AB}$.

26) Find AD. (Use either method.)

27) Find AB.

28) Find the midpoint of $\overline{CB}$.

29) Find BC.

TEST PRACTICE

MULTIPLE-CHOICE QUESTIONS: Write the letter of the correct answer.

1) Use the diagram to choose the correct answer. $l \parallel g$

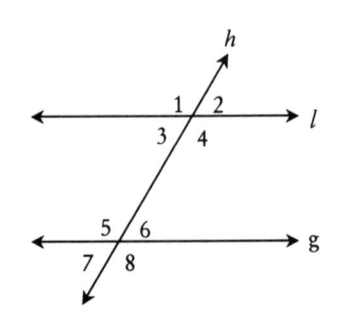

   a. m∠1 = m∠6
   b. m∠3 + m∠2 = 90°
   c. m∠4 + m∠6 = 180°
   d. m∠2 = m∠8

Meeting the Challenge...Mathematics

2) If two parallel lines are cut by a transversal, the corresponding angles are
   a. complementary   b. supplementary   c. vertical   d. equal

3) Find the length of the segment with endpoints at A(1, 4) and B(3, 1).
   a. 3.6   b. 5   c. 6   d. 3.8

4) Which lines are definitely parallel?
   a. 112°/58°   b. 100°/100°   c. 55°/115°   d. 50°/50°

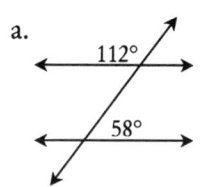

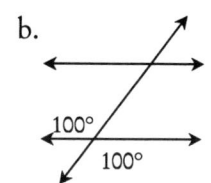

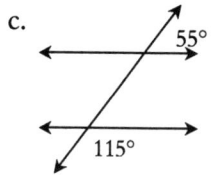

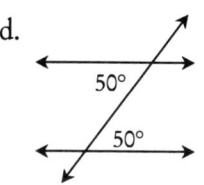

5) Find the angle measures.

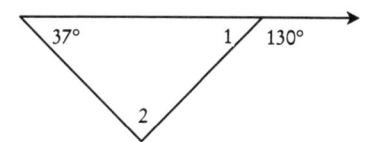

   (triangle with 37°, ∠1, 130°, ∠2)

   a. m∠1 = 50° and m∠2 = 49°   c. m∠1 = 50° and m∠2 = 93°
   b. m∠1 = 37° and m∠2 = 106°   d. m∠1 = 59° and m∠2 = 84°

SHORT ANSWER QUESTIONS:

6) Find the angle measures. $m \parallel n$

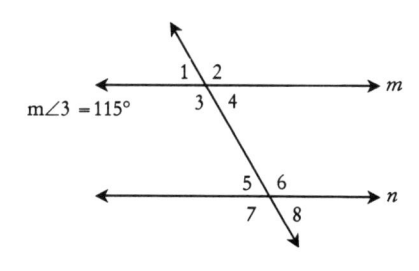

   (diagram with m∠3 = 115°, angles 1,2,3,4 on line m and 5,6,7,8 on line n)

   m∠1 = _____   m∠6 = _____
   m∠2 = _____   m∠7 = _____
   m∠4 = _____   m∠8 = _____
   m∠5 = _____

7) To answer the following questions, refer to the angle numbers in diagram #6.

   a) Name two pairs of corresponding angles.
   _____ _____

   b) Name one pair of vertical angles. _____

   c) Name one pair of alternate interior angles. _____

   d) Name one pair of alternate exterior angles. _____

EXTENDED RESPONSE QUESTIONS:

8) $\overline{AB} \parallel \overline{CD}$; $\overline{BC} \parallel \overline{DE}$ a) m∠1 = _____

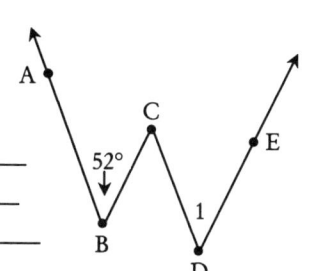

   b) How did you determine m∠1? _____
   c) Would your answer be the same if one pair of lines was not parallel? _____
   d) Explain. _____

   (diagram showing W-shape with points A, B, C, D, E; 52° at B; ∠1 at D)

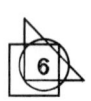

9) Find the missing angle measures. Give a short hint you used to find each measure. There can be more than one way to solve for a measure. $a \parallel b$

Angle Measure   Hint

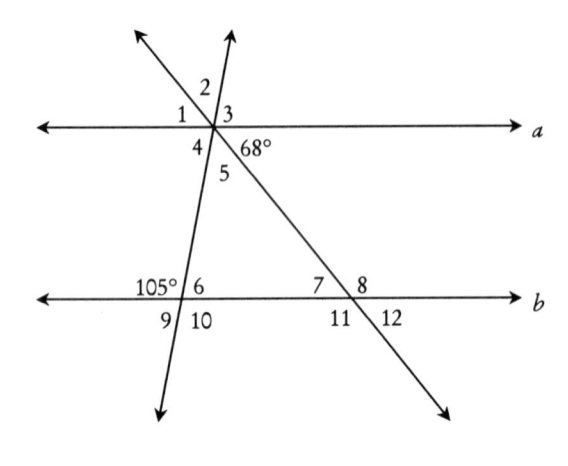

m∠1 = _____ _____
m∠2 = _____ _____
m∠3 = _____ _____
m∠4 = _____ _____
m∠5 = _____ _____
m∠6 = _____ _____
m∠7 = _____ _____
m∠8 = _____ _____
m∠9 = _____ _____
m∠10 = _____ _____
m∠11 = _____ _____
m∠12 = _____ _____

10) Looking at the map, we can assume that Main Street and Spring Street are parallel to each other.

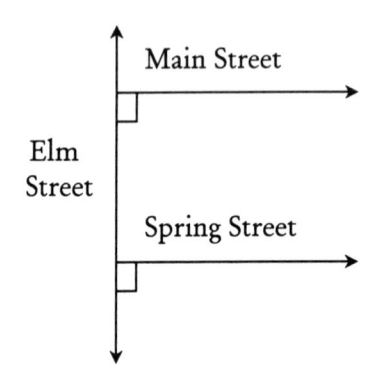

Justify the above statement. It is important to use key words relating to lines and angles.

Explanation:

CLOSURE

These are some of the real-life applications involving lines and angles that I encountered while running errands one Saturday morning.
- Perpendicular lines and right angles were primarily involved when I turned corners. I did notice that one intersection was more of a 'Y'. That was an acute angle.
- I pulled into a parking space that was perpendicular to the driving lane. Later I pulled into a space that was a turn like an acute angle. The second space was easier to pull into than the first, but it was more difficult to see when backing out.
- One of the tables at the grocery store was supported by scissor legs that looked like intersecting lines forming vertical angles. Those legs combined with the tabletop and with the floor reminded me of parallel lines (table and floor) cut by transversals (scissor legs).

Meeting the Challenge...Mathematics

CONGRUENT AND SIMILAR FIGURES

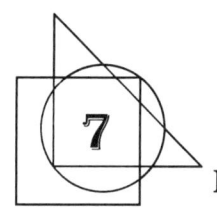

7

Recognize and apply characteristics of congruent and similar figures.

FOCUS: Let's look at mathematics in our own back yard.

There is a very tall tree on the corner of Mr. King's property. He is quite concerned that it could fall on his house during a bad storm. The tree is 55 feet from the house but he has no idea how tall it is.

One day as Mr. King stands in his yard, he notices that his 6-foot frame casts a four foot shadow. He immediately measures the shadow of the tree and finds that it measures 45 feet. Use your knowledge of similar figures to help Mr. King find the height of the tree. Does Mr. King need to worry about the tree falling on his house?

PURPOSE: When you have completed this lesson, you will be able to use the concepts of congruency and similarity to find missing parts in geometric figures and to solve application problems.

WHAT YOU NEED TO KNOW

MATHEMATICS AND PROBLEM SOLVING VOCABULARY:

1. Altitude
2. Bisector
3. Congruent
4. Corresponding parts
5. Equilateral triangle
6. Hypotenuse of a right triangle
7. Isosceles triangle
8. Parallelogram
9. Polygon
10. Postulate
11. Proportion
12. Quadrilateral
13. Ratio
14. Rhombus
15. Right triangle
16. Scale, ratio or factor
17. Scalene triangle
18. Similar figures
19. Square
20. Theorem
21. Trapezoid

MEMORIZE THESE FACTS AND FORMULAS:

1. The symbol that tells two figures are congruent is ≅.
2. Figures are said to be congruent when they have the same size and shape.
3. A symbol that tells sides or angles are corresponding is ↔.
4. Corresponding parts of congruent figures are equal.
5. The altitude (height) of a triangle is drawn from the vertex of an angle to the opposite side where it forms perpendicular lines with that side.
6. An included angle is the angle (∠A) formed from two sides ($\overline{AB}$ and $\overline{AC}$).
7. An included side is the side ($\overline{AC}$) between two angles (∠A and ∠C).
8. Triangles are classified by their sides and angles.

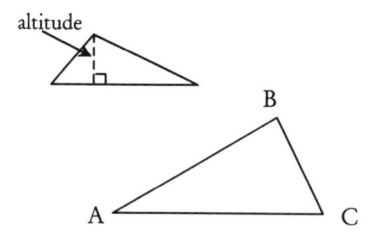

a. If classified by sides, there are three types of triangles.	b. If classified by angles, there are three types of triangles.
▪ A scalene triangle has no equal sides. ▪ An isosceles triangle has two equal sides referred to as legs. ▪ An equilateral triangle has all three sides equal.	▪ An obtuse triangle has an obtuse angle within it. ▪ A right triangle has a right angle within it. ▪ An acute triangle has three acute angles.

© 2003 Orange Frazer Press. All Rights Reserved.

9. Look at the right triangle.
   a. The side opposite the right angle is the hypotenuse.
   b. The legs form the right angle.

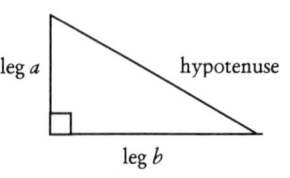

10. Look at the isosceles triangle. The legs are equal. The base angles are equal.
    a. If two angles in a triangle are equal, the sides opposite those angles are equal.
    b. If two sides in a triangle are equal, the angles opposite those sides are equal.
    c. The altitude from the vertex angle of an isosceles triangle bisects the vertex angle and the base.

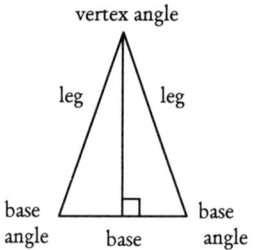

11. In an equilateral or equiangular triangle, the measures of all three sides are equal and all three angles are equal.

Establish Congruent Triangles

1. To assume congruent triangles, it is not necessary to prove that all six pairs of corresponding parts are equal. The following rules (postulates and theorems) will establish congruent triangles. The triangles are congruent if it can be established that three special pairs of corresponding parts (not three angles) are equal.

   Memorize these rules.
   a. SSS postulate   three sides of both triangles
   b. SAS postulate   two sides and the included angle of both triangles
   c. ASA postulate   two angles and the included side of both triangles
   d. AAS theorem   two angles and a non-included side of both triangles

2. Two right triangles are congruent if the following corresponding parts are equal.
   a. HL   the hypotenuse and a leg of both right triangles
   b. LL   two legs of both right triangles
   c. HA   the hypotenuse and an acute angle of both right triangles
   d. LA   a leg and an angle of both right triangles

Properties of Quadrilaterals

1. A parallelogram is a quadrilateral with both pairs of opposite sides parallel. Learn these properties of parallelograms.
   a. Opposite sides are equal.
   b. Opposite angles are equal.
   c. The diagonals of a parallelogram bisect each other.
   d. The sum of the interior angles of a parallelogram is 360°.

2. Special parallelograms:
   a. A rectangle is a parallelogram with four right angles.
   b. A rhombus is a parallelogram with four equal sides.
   c. A square is a parallelogram with four right angles and four equal sides.

Meeting the Challenge...Mathematics

3. A trapezoid is a quadrilateral with one pair of parallel sides (bases).

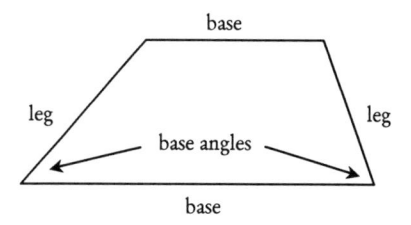

4. An isosceles trapezoid has the following properties:
    a. parallel bases
    b. equal legs and equal base angles
    c. parallel median and bases
    d. length of the median = one-half the sum of the bases

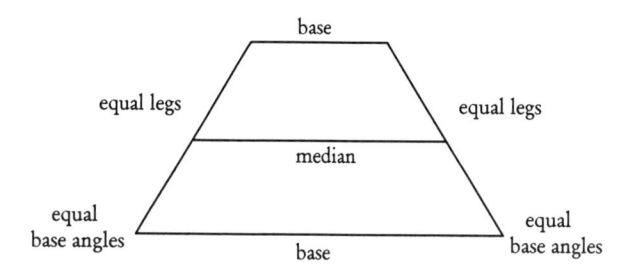

Similar Figures

1. The symbol that tells two figures are similar is ~.
2. Figures are said to be similar if their corresponding angles are equal and their corresponding sides are in proportion.
3. In similar figures, the ratio of two corresponding sides is called the scale or ratio.
4. In similar figures, the perimeters have the same ratio as the sides.

Similar Triangles

1. If two angles of one triangle are equal to two angles of another triangle, the triangles are similar.
2. In two similar triangles, the corresponding angles are equal, while corresponding sides, altitudes and perimeters are in proportion.
3. If a line or segment is parallel to one side of a triangle ($\overline{MO} \parallel \overline{BC}$) and intersects the other two sides, the following ideas are true.
    a. The line creates a smaller similar triangle ($\triangle AMO$) contained within the original triangle ($\triangle ABC$).
    b. The line divides those sides of the triangle proportionally.

$$\frac{AM}{MB} = \frac{AO}{OC}$$

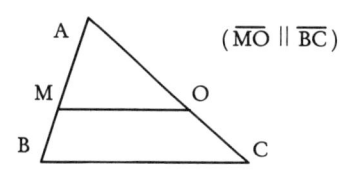

BE ABLE TO PERFORM THESE OPERATIONS:

Identifying Congruent Figures

1. Establish that two triangles are congruent by naming the rules (postulates and theorems) that are appropriate.
   Example:

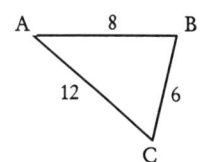

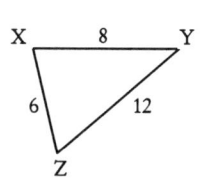

   Since all three pairs of corresponding sides are equal, we can say that $\triangle ABC \cong \triangle YXZ$.

   Justification: If three sides of one triangle are congruent to three sides of another triangle, the triangles are congruent. (SSS)

2. For each example:
   a. Determine if the two figures are congruent.
   b. Show their correspondence by naming them using the corresponding vertices and marking the corresponding parts on the diagram.
   c. List their corresponding parts.
   Example 1: Compare shapes, angles and sides.
   Example 2: Use rotation, then compare shapes, angles and sides.

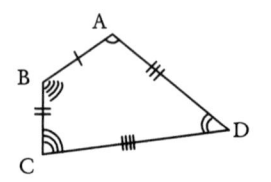

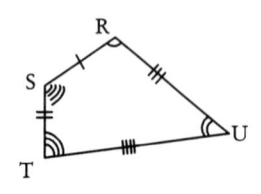

   a. Yes, corresponding angles and sides are equal.
   b. Quadrilateral ABCD ≅ Quadrilateral RSTU
   c. Corresponding parts:

   $\angle A \leftrightarrow \angle R$ $\overline{AB} \leftrightarrow \overline{RS}$
   $\angle B \leftrightarrow \angle S$ $\overline{BC} \leftrightarrow \overline{ST}$
   $\angle C \leftrightarrow \angle T$ $\overline{CD} \leftrightarrow \overline{TU}$
   $\angle D \leftrightarrow \angle U$ $\overline{DA} \leftrightarrow \overline{UR}$

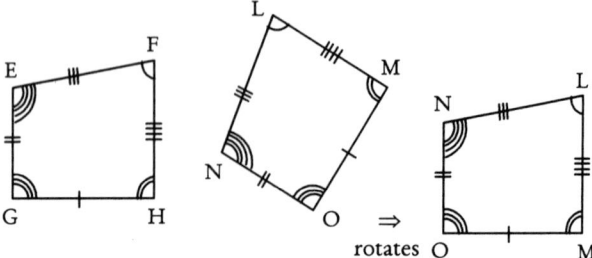

   a. Yes, corresponding angles and sides are equal.
   b. Quadrilateral EFHG ≅ Quadrilateral NLMO
   c. Corresponding parts:
   $\angle E \leftrightarrow \angle N$ $\overline{EF} \leftrightarrow \overline{NL}$
   $\angle F \leftrightarrow \angle L$ $\overline{FH} \leftrightarrow \overline{LM}$
   $\angle H \leftrightarrow \angle M$ $\overline{HG} \leftrightarrow \overline{MO}$
   $\angle G \leftrightarrow \angle O$ $\overline{GE} \leftrightarrow \overline{ON}$

Identifying Similar Figures

1. Two figures are similar if the corresponding angles are equal and the corresponding sides are in proportion.
   Example: Verify that these quadrilaterals are similar.

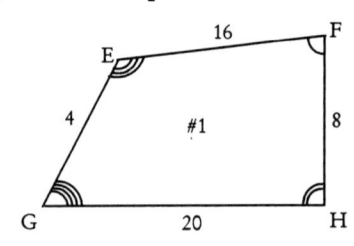

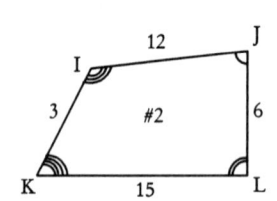

   Solution: We know that corresponding angles are equal by the markings on the diagram.
   We need to verify that corresponding sides are in proportion.

   Quadrilateral #1: $\dfrac{EF}{IJ} = \dfrac{FH}{JL} = \dfrac{HG}{LK} = \dfrac{GE}{KI}$ With substitutions: $\dfrac{16}{12} = \dfrac{8}{6} = \dfrac{20}{15} = \dfrac{4}{3}$

   To verify that the ratios form a proportion, simplify each ratio or use cross products with each pair.
   Each ratio simplifies to $\dfrac{4}{3}$. The quadrilaterals are similar.

2. Use proportions to solve for the missing sides in these similar triangles.
   Example:

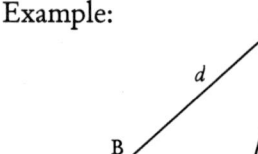

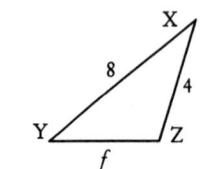

   To solve for d: $\dfrac{4}{6} = \dfrac{8}{d}$, $4d = 48$ $d = 12$

   To solve for f: $\dfrac{4}{6} = \dfrac{f}{3}$, $6f = 12$ $f = 2$

Meeting the Challenge...Mathematics

Using Similar Triangles in Scale Drawings

To construct a scale model or drawing, a ratio is used to set up proportions that relate the original measures to the scaled down version. In Lesson #3, the ratio or scale factor was given. In this lesson we are working in the world of geometry. We can solve for the scale factor by identifying corresponding sides and constructing a ratio.

Example: On a map, the distance from Dayton to Athens measures 9 inches. This distance represents 171 miles. Find the scale ratio. Create a proportion to find the number of miles that $4\frac{3}{4}$ inches represents.

Solution: Find the scale ratio by reducing. $\frac{9}{171} = \frac{3}{57} = \frac{1}{19}$

Create a proportion relating the corresponding parts. Proportion: $\frac{1}{19} = \frac{4.75}{m}$ Answer: 90.25 miles

BE ABLE TO PERFORM THESE CALCULATOR OPERATIONS:

This objective includes some calculations using proportions. Use your calculator as needed.

GUIDED PRACTICE

A) Use the figures to complete the statements below.

1) 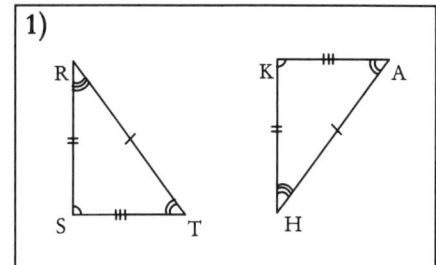
a) $\triangle STR \cong \triangle$ _____
b) $\triangle TRS \cong \triangle$ _____
c) $\triangle RTS \cong \triangle$ _____

2) Pentagon I $\cong$ Pentagon II 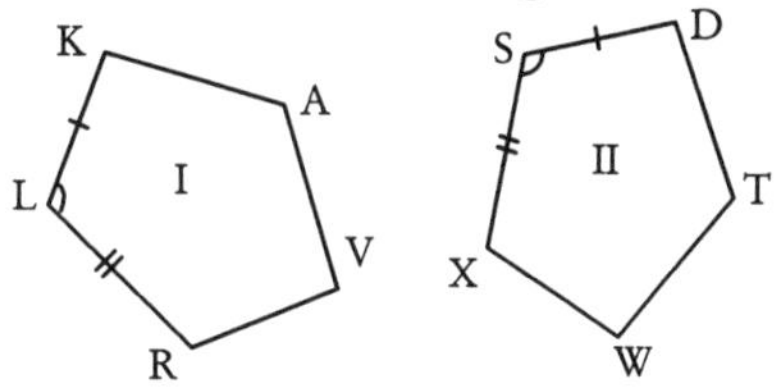
a) ⌂KLRVA $\cong$ ⌂ _____
b) $\angle K \cong$ _____   c) $\overline{RV} \cong$ _____
   $\angle L \cong$ _____ $\overline{VA} \cong$ _____
   $\angle R \cong$ _____ $\overline{AK} \cong$ _____
   $\angle V \cong$ _____ $\overline{KL} \cong$ _____
   $\angle A \cong$ _____ $\overline{LR} \cong$ _____

3) 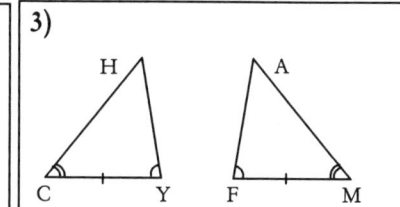
a) $\triangle HCY \cong \triangle$ _____
b) $\overline{HC} \cong$ _____   c) $\angle H \cong$ _____
   $\overline{HY} \cong$ _____ $\angle Y \cong$ _____
   $\overline{CY} \cong$ _____ $\angle C \cong$ _____

B) Find the measures of the missing sides and angles.

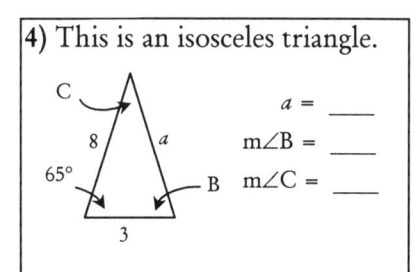

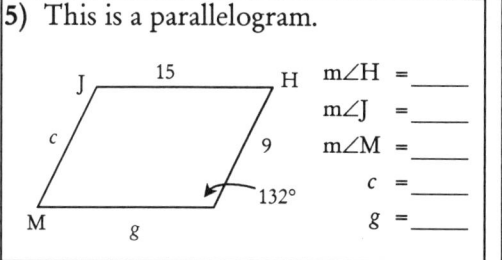

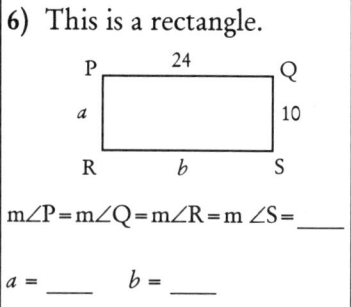

4) This is an isosceles triangle.
a = _____
m∠B = _____
m∠C = _____

5) This is a parallelogram.
m∠H = _____
m∠J = _____
m∠M = _____
c = _____
g = _____

6) This is a rectangle.
m∠P = m∠Q = m∠R = m∠S = _____
a = _____   b = _____

Meeting the Challenge...Mathematics

C) Use these similar figures to answer the questions.

7) △ABC ~ △DEF

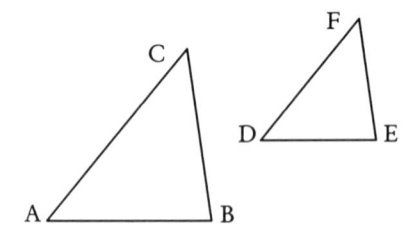

a) Give the corresponding sides.
  $\overline{AB}$ corresponds to _____.
  $\overline{BC}$ corresponds to _____.
  $\overline{AC}$ corresponds to _____.
b) Fill in the extended proportion.

$$\frac{AB}{e} = \frac{BC}{f} = \frac{AC}{g}$$

$e = $ _____
$f = $ _____
$g = $ _____

8) Quadrilateral ABDC ~ Quadrilateral LMKN

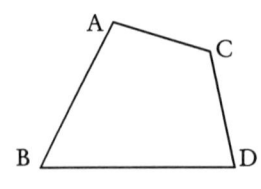

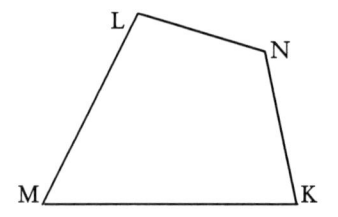

a) Give the corresponding sides.
  $\overline{AB}$ corresponds to _____.
  ___ corresponds to _____.
  ___ corresponds to _____.
  ___ corresponds to _____.
b) Fill in the extended proportion.

$$\frac{AB}{p} = \frac{BD}{q} = \frac{x}{t} = \frac{y}{v}$$

$x = $ _____ $q = $ _____
$y = $ _____ $t = $ _____
$p = $ _____ $v = $ _____

D) In these similar figures, set up the proportion and find the missing values.

9) Quadrilateral A ~ Quadrilateral B

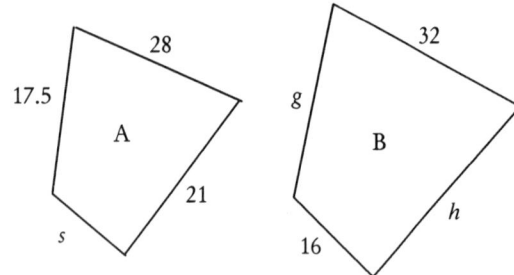

(A: 28, 17.5, 21, s; B: 32, g, h, 16)

Solve for:	Proportions	Answers
s	_____	___
g	_____	___
h	_____	___

10) Polygon I ~ Polygon II

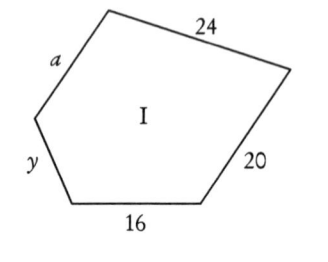

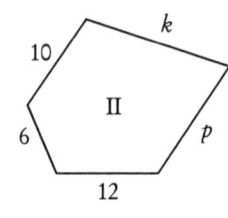

(I: 24, a, y, 16, 20; II: k, 10, 6, 12, p)

Solve for:	Proportions	Answers
a	_____	___
y	_____	___
k	_____	___
p	_____	___

Meeting the Challenge...Mathematics

E) Tell if the two triangles are similar and justify. Find the missing angle measures.

11) Are these triangles similar?   Yes   No

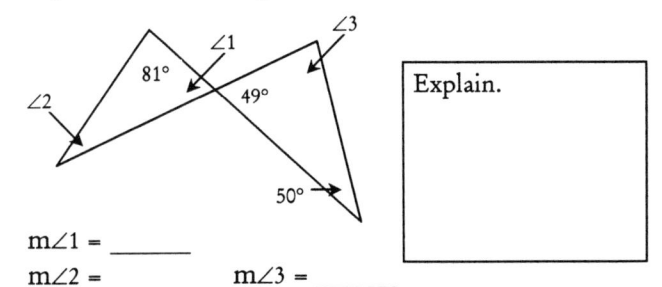

Explain.

m∠1 = _____
m∠2 = _____   m∠3 = _____

12) Are these triangles similar?   Yes   No

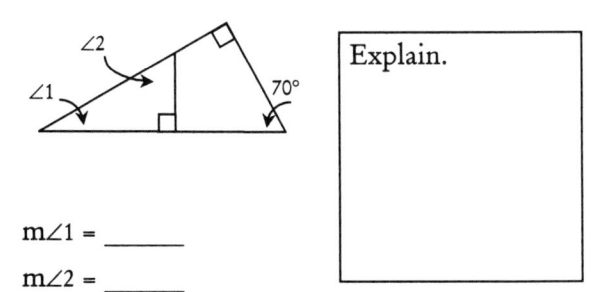

Explain.

m∠1 = _____
m∠2 = _____

F) Use the scale drawings to find the missing values.

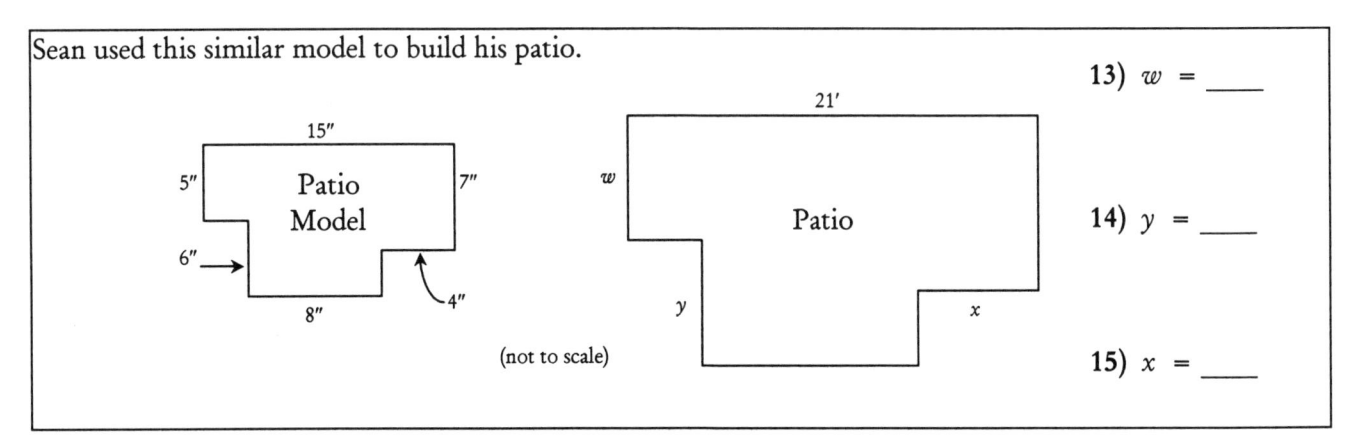

Sean used this similar model to build his patio.

(not to scale)

13) w = ____

14) y = ____

15) x = ____

INDEPENDENT PRACTICE

A) Answer the questions regarding the following congruent figures.

1) Which angles and sides are corresponding?

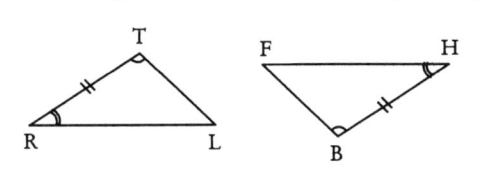

a) △HBF ≅ △ ____ using ____ rule.

b) $\overline{BF}$ ≅ ____
   $\overline{HB}$ ≅ ____
   $\overline{FH}$ ≅ ____

c) ∠H ≅ ____
   ∠B ≅ ____
   ∠F ≅ ____

2) These figures are congruent. Complete these statements.

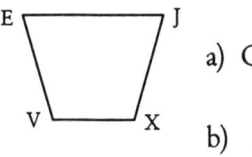

a) Quad. VXJE ≅ Quad. _____

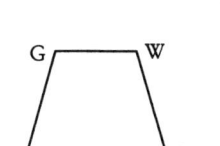

b) ∠G ≅ ____
   ∠A ≅ ____
   ∠W ≅ ____
   ∠Q ≅ ____

c) $\overline{GW}$ ≅ ____
   $\overline{GA}$ ≅ ____
   $\overline{AQ}$ ≅ ____
   $\overline{QW}$ ≅ ____

© 2003 Orange Frazer Press. All Rights Reserved.

B) The following pairs of triangles are congruent. Find the missing sides and angles.

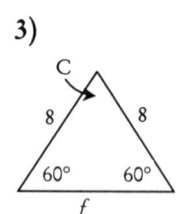

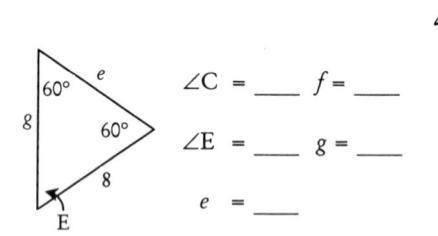

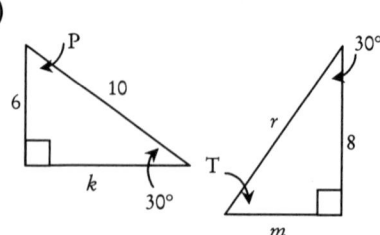

3)

∠C = ___ f = ___
∠E = ___ g = ___
e = ___

4)

k = ___
∠P = ___
∠T = ___
m = ___
r = ___

C) The congruent parts are shown by the tick marks. Name the additional side or angle necessary to prove the triangles are congruent.

5) In order to use SSS to establish that these triangles are congruent, what additional pair of corresponding parts must be equal?

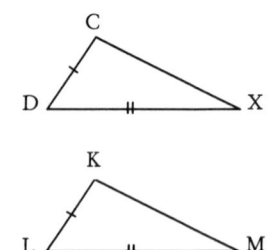

_____

6) In order to use SAS to establish that these triangles are congruent, what additional pair of corresponding parts must be equal?

_____

D) The corresponding sides of two triangles are shown in order. Use these sides to determine if the triangles are similar. Show proportions.

7) The sides of Δ I (10, 15, 20) correspond to the sides of Δ II (8, 12, 16).

   a) Are these triangles similar?   Yes   No

   b) Give proportions:

8) The sides of Δ I (3.6, 7.2, 9) correspond to the sides of Δ II (5.4, 10.8, 13.5).

   a) Are these triangles similar?   Yes   No

   b) Give proportions:

Meeting the Challenge...Mathematics

E) Find the missing values.

These are similar figures, so we know that corresponding angles are equal.

The given values are:
LX = 12, LK = 8, KE = 10, EY = 18, YX = 16, RP = 9

Set up a proportion or expression to solve for these values.

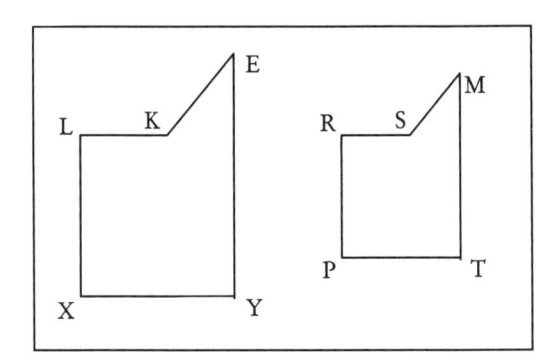

9) RS
Answer: _____

10) SM
Answer: _____

11) MT
Answer: _____

12) TP
Answer: _____

13) perimeter of LKEYX
Answer: _____

14) perimeter of RSMTP
Answer: _____

F) Solve these similar triangle problems.

List the corresponding angles and sides and mark them on the diagram.

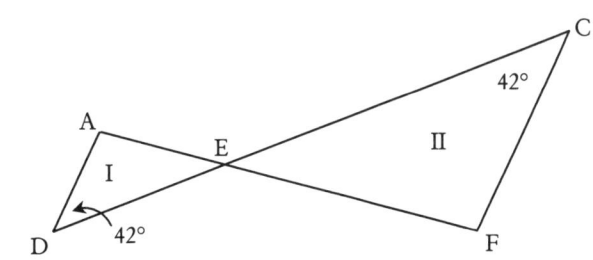

15) corresponding angles
a) _____
b) _____
c) _____

16) corresponding sides
a) _____
b) _____
c) _____

17) We know the following:
△I ~ △II
AE = 9"   EF = 20"   AD = 7"

Solve for FC.   FC = _____

18) The basic way to verify that two triangles are similar is to establish that two pairs of corresponding angles are equal. In this problem it is given that m∠E = m∠K.

    a) _____ and _____ are another pair of equal corresponding angles.

    b) Why are they equal? _____

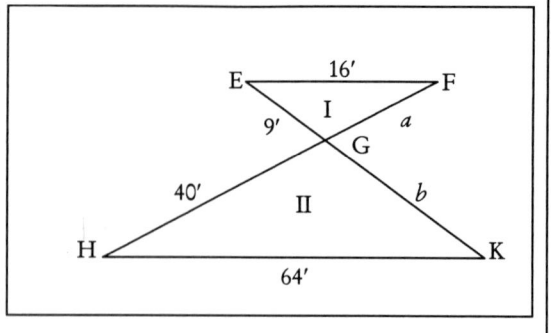

19) Give the proportion and solve for side *a*.

_____   *a* = ___

20) Give the proportion and solve for side *b*.

_____   *b* = ___

TEST PRACTICE

MULTIPLE-CHOICE QUESTIONS:

1. If △ABC ~ △XYZ, BC = 8 and YZ = 6, the length of another pair of corresponding sides could be ____

   a. 12 and 10   b. 4 and 5   c. 12 and 9   d. 3 and 4

2. The scale on a map is $\frac{3}{4}$ inch = 150 miles. Find the distance between cities that are $2\frac{1}{2}$ inches apart. ____

   a. 225 miles   b. 425 miles   c. 500 miles   d. 300 miles

3. If △RST ≅ △GFE, choose the correct pair of corresponding parts. ____

   a. ∠E ↔ ∠R   b. TS ↔ EF   c. RS ↔ FE   d. ∠S ↔ ∠E

4. △ACB is similar to △KML. Find the length of $\overline{LM}$. ____

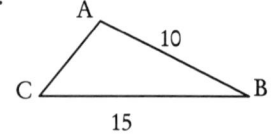

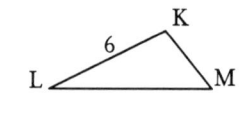

   a. 9   b. 11   c. 12   d. 10

5. If △PAM ~ △DOT, which of these is a correct proportion? ____

   a. $\frac{AM}{DO} = \frac{PA}{TD}$ b. $\frac{MA}{TO} = \frac{PM}{DO}$ c. $\frac{OT}{AM} = \frac{DO}{PA}$ d. $\frac{TD}{MP} = \frac{TO}{AP}$

Meeting the Challenge...Mathematics

SHORT ANSWER QUESTIONS:

6) At the same time of day, a tree casts a shadow 30 feet long and a mailbox nearby casts a shadow 3 feet long. The mailbox is 4 feet high. Find the height of the tree.

a) Set up a proportion to find the height of the tree.

b) Solve for the height of the tree.

Answer: _____

7) Identify the congruency rule (SSS, SAS, AAS or ASA) that applies below.

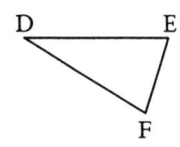

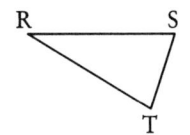

a) $\angle D = \angle R$; $\overline{DE} = \overline{RS}$; $\angle E = \angle S$ Rule: _____

b) $\overline{RT} = \overline{DF}$; $\angle T = \angle F$; $\overline{TS} = \overline{FE}$ Rule: _____

c) $\overline{DF} = \overline{RT}$; $\overline{DE} = \overline{RS}$; $\overline{EF} = \overline{ST}$ Rule: _____

EXTENDED RESPONSE QUESTIONS:

The two triangles are congruent and $\overline{GX} \parallel \overline{YH}$.

8) Because these lines are parallel, what else can you determine?

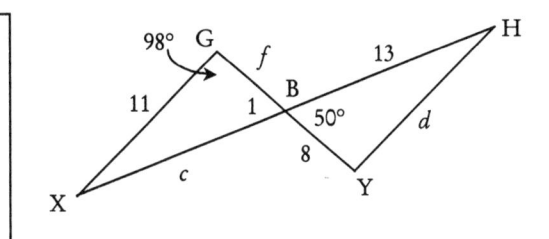

9) Find the missing angles/sides and explain your reason for each solution.

∠1 = ____ Explanation: _____

∠Y = ____ Explanation: _____

∠X = ____ Explanation: _____

f = ____ Explanation: _____

c = ____ Explanation: _____

d = ____ Explanation: _____

10) A ramp 15 feet long is used to load and unload boxes from a delivery truck. After going 6 feet up the ramp, the distance to the ground is 2 feet. How far above the ground is the upper end of the ramp?

   a) Make a sketch and verify that these are similar triangles.
   _____
   _____

   b) Give a proportion to solve this problem.

   c) Solve.

   Answer: _____

   Sketch and label the dimensions.

CLOSURE

The shadows of Mr. King and the tree allow us to use similar triangles. The sketch shows the correspondence between Mr. King and his shadow with the tree and its shadow. These two triangles are similar.

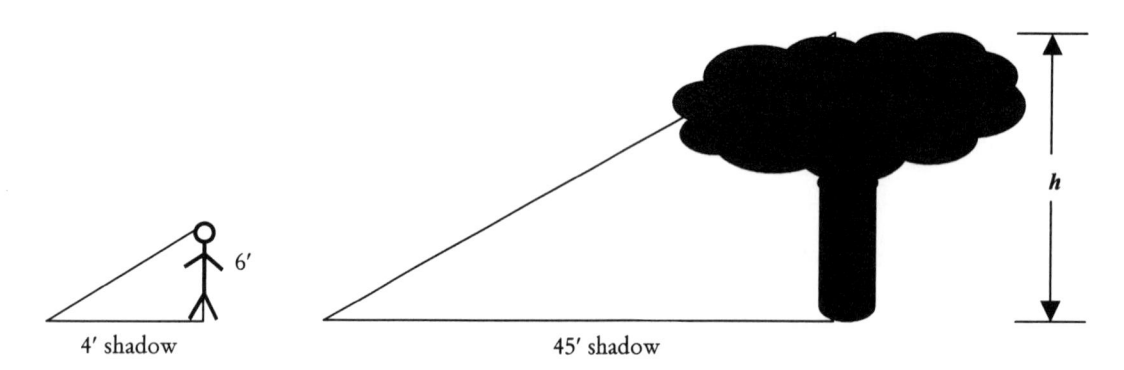

6′

4′ shadow   45′ shadow   h

Use a proportion to solve for the height of the tree. $\frac{6}{4} = \frac{h}{45}$ $4h = 270$ $h = 67.5$ feet

If the tree falls in the direction of the house, it *will* fall on the house.

Meeting the Challenge...Mathematics

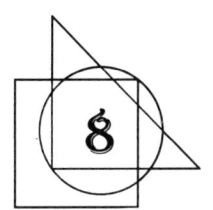

GEOMETRIC FIGURES and OBJECTS

Visualize, recognize, classify and apply properties of
two-dimensional figures and three-dimensional objects.

FOCUS: Geometric figures can be found in nearly every aspect of our lives. Let's discover more about a tile pattern that is particularly fascinating.

To make some extra money, Calvin helped his father install new vinyl flooring in their kitchen. Calvin was quite surprised to find that many designs included geometric figures and patterns. He was particularly fascinated with an interlocking pattern of six-sided figures. When he studied a single figure, he noticed that all six sides looked to be the same length and all six angles appeared to have the same measure.
1. Name the figure and tell all you know about it.
2. Using Calvin's observations, sketch the figure and the interlocking pattern.
3. Give an example of this same interlocking design in the world of nature.

PURPOSE: When you have completed this lesson, you will be able to apply properties and characteristics of geometric figures and objects.

WHAT YOU NEED TO KNOW

MATHEMATICS AND PROBLEM SOLVING VOCABULARY:

1. Central angle
2. Chord
3. Circle
4. Circumference
5. Diameter
6. Equilateral triangle
7. Hexagon
8. Isosceles triangle
9. Lateral edge
10. Lateral face
11. Line of reflection or symmetry
12. Major arc
13. Minor arc
14. Median
15. Octagon
16. Parallelogram
17. Pentagon
18. Polygon
19. Prism
20. Quadrilateral
21. Radius
22. Reflection
23. Regular polygon
24. Rhombus
25. Rotation
26. Scalene triangle
27. Semicircle
28. Slant height
29. Square
30. Tangent line
31. Transformation
32. Translation
33. Trapezoid

MEMORIZE THESE FACTS AND FORMULAS:

Properties of Triangles

1. Associate these properties of sides and angles with the appropriate triangles.

Equilateral/Equiangular △		Isosceles △		Scalene △	
congruent sides	congruent angles	congruent sides	congruent base angles	congruent sides	congruent angles
3	3	2	2	0	0

© 2003 Orange Frazer Press. All Rights Reserved.

2. Triangles can be classified by their angles.

Acute Triangle	Obtuse Triangle	Right Triangle
three acute angles	one obtuse angle	one right angle |//

Properties of Quadrilaterals

Associate these properties with the appropriate quadrilateral(s).

Properties of Special Quadrilaterals	Parallelogram	Rectangle	Rhombus	Square
1) Opposite sides are parallel.	x	x	x	x
2) Opposite sides are congruent.	x	x	x	x
3) Opposite angles are congruent.	x	x	x	x
4) Diagonals form congruent triangles.	x	x	x	x
5) Diagonals bisect each other.	x	x	x	x
6) Diagonals are congruent.		x		x
7) Diagonals are perpendicular.			x	x
8) Diagonals bisect opposite angles.			x	x
9) All angles are right angles.		x		x
10) All sides are congruent.			x	x

Properties of More Special Quadrilaterals	Trapezoid	Isosceles Trapezoid
1) Bases are parallel.	x	x
2) Legs are congruent.		x
3) Base angles are congruent.		x
4) Diagonals are congruent.		x
5) The median is parallel to the bases.	x	x
6) The length of the median is one-half the sum of the bases.	x	x

Properties of Polygons

1. Polygons are classified by the number of sides in the figure.

Polygon	Triangle	Quadrilateral	Pentagon	Hexagon	Octagon	Decagon
Number of sides	3	4	5	6	8	10
Shape						

2. A regular polygon is both equiangular and equilateral.
3. At any given vertex of a polygon, the interior and exterior angles are supplementary.

4. There are two methods that can be used to find the total degrees in the interior of a polygon.
   a. Triangle method: Pick one vertex in the polygon and draw a diagonal to every nonadjacent vertex to form triangles. Count the number of triangles and multiply by 180°.
   b. Formula method: $(n - 2) \times 180°$ will find the total number of interior degrees in a polygon with n sides.
5. The sum of the measures of the exterior angles of a polygon is 360°.

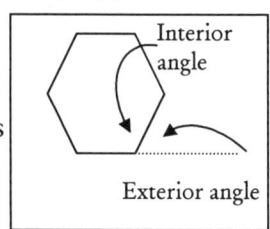

Properties of Circles

1. To name a circle, use the circle symbol and the letter naming the center point.
   Symbol: ⊙ O   Read: circle O
2. A chord is a segment through the interior of a circle with endpoints on the circle.
3. The radius ($\overline{OA}$) is the distance from the center to the circumference.
4. The diameter ($\overline{BC}$) is a chord through the center of the circle.
5. The length of the diameter is twice the length of the radius.
6. A tangent line intersects the circle in exactly one point. $\overleftrightarrow{k}$ is a tangent line.
7. P is the point of tangency where $\overleftrightarrow{k}$ touches the circle.
8. Central angles (∠1 & ∠2) are formed where two radii intersect in the center.
9. The minor arc of ∠1 is $\overset{\frown}{AB}$ (less than a semicircle). The major arc is $\overset{\frown}{ACB}$ (greater than a semicircle).
10. There are 360° in a circle and 180° in a semicircle.

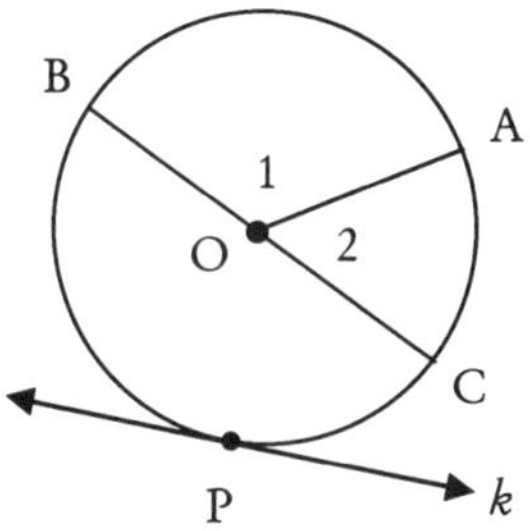

Transformations

A geometric transformation takes an original figure and maps its shape and size to a second figure (image) by using reflection, rotation or translation.

1. **Reflection**   A reflection creates a mirror image of the original figure. The 'mirror' line is called the line of symmetry.

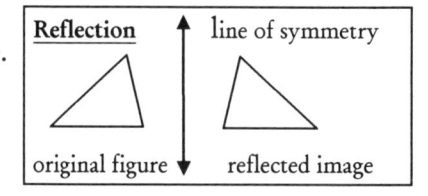

2. **Rotation**   Every point on the figure turns (rotates) a certain number of degrees. A counterclockwise rotation is positive and a clockwise rotation is negative.

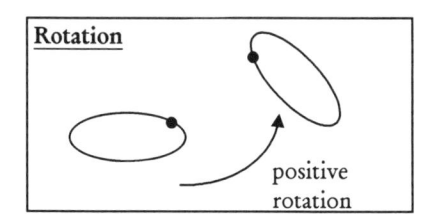

3. **Translation**   During a translation all the points of the figure slide an equal distance in the same direction.

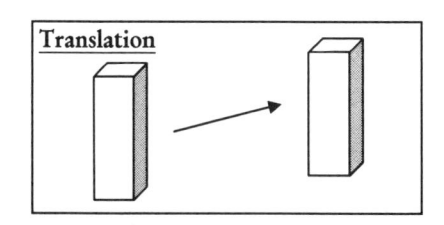

Three-Dimensional Objects

Rectangular Solid	Cube	Cone	Cylinder	Prism	Pyramid	Sphere
a) This object is a prism with a rectangular base. b) There are three pairs of rectangles that make up the six faces.	a) This object is a prism with a square base. b) There are six congruent square faces.	a) The base is a circle. b) The altitude is the height of the cone. c) The slant height connects the vertex to any point on the edge of the base.	a) The bases are circles. b) The altitude is the height of the cylinder.	a) The bases are congruent polygons. b) The altitude is the height of the prism. c) Lateral faces are the sides that are not bases.	a) The base is a polygon. b) The lateral faces are triangular sides. c) The height or altitude of a lateral face is the slant height.	A sphere is the set of all points a fixed distance from a given point called the center. A sphere is a 3-D circle.

BE ABLE TO PERFORM THESE OPERATIONS:

Identifying Figures

Become familiar with the properties of two-dimensional figures and three-dimensional objects.
Be able to identify figures from their properties and also properties from their figures.
Example: Name a quadrilateral with only one pair of opposite sides parallel.   Answer: trapezoid

Classifying Figures

Be able to classify and find the common characteristics of two-dimensional figures and three-dimensional objects.
Example: Name two parallelograms that have equal sides.   Answer: rhombus and square

Finding angle measures in figures

1. **Problem situation for polygons:** Use the diagonal/triangle method to find the measures of interior and exterior angles of a polygon.
    Example: Find the measure of the interior and exterior angles of a regular pentagon.
    Solution: Find the interior angle measure, then find the exterior angle measure.
    a) Find the total number of interior degrees in the pentagon using the diagonal/triangle method. There are 540° (180° × 3) in the interior of the pentagon. Find the measure of an interior angle by dividing the total number of degrees in the polygon (540°) by the number of angles (5) in the figure. The measure of an interior angle is 108° (540° ÷ 5).
    b) Find the measure of an exterior angle by dividing 360° by the number of exterior angles (5). The measure of an exterior angle is 72°.
    c) These answers can be verified because we know an interior angle and an exterior angle are supplementary.   (108° + 72° = 180°)

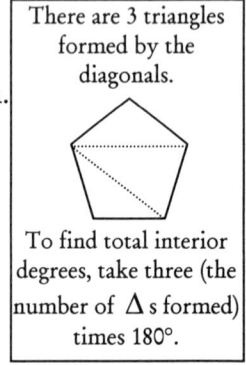

> There are 3 triangles formed by the diagonals.
>
> To find total interior degrees, take three (the number of Δs formed) times 180°.

Meeting the Challenge...Mathematics

2. Problem situation for circles: Find the missing measures of central angles.
   Example: Find the measure of ∠1.
   Solution: There are 360° in a circle. Since $\overline{BC}$ is a diameter, it divides the circle into two semicircles of 180° each.
   Equation: m∠1 + 68° = 180°
   Answer: m∠1 = 112°

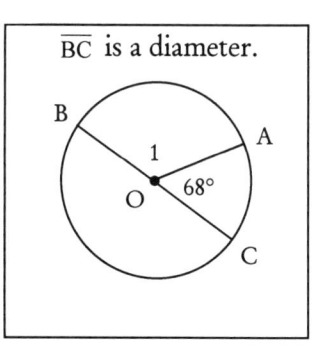

$\overline{BC}$ is a diameter.

Visualizing and Applying Transformations

Use reflection, rotation and translation to create images and move them.
Example: On the graph, duplicate the given triangle and move it three units to the right. Give the coordinates of the new vertices.

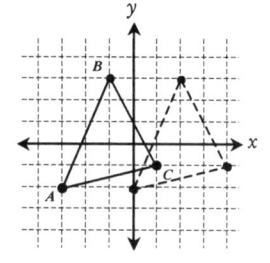

Answer: Point A translates to (0, −2).
Point B translates to (2, 3).
Point C translates to (4, −1).

BE ABLE TO PERFORM THESE CALCULATOR OPERATIONS:

This objective involves a minimum number of calculations, however it might be necessary to use the distance formula or the Pythagorean theorem to find lengths of sides.

GUIDED PRACTICE

A) Name the polygons.

1) 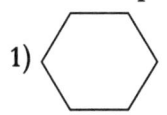 [hexagon]   2) 4-sided polygon   3) 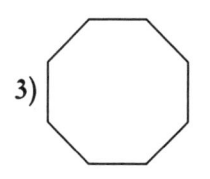 [octagon]   4) 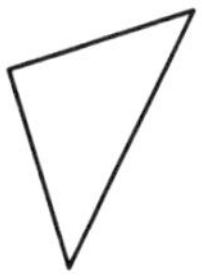 [triangle]   5) 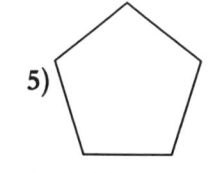 [pentagon]   6) 10-sided polygon

Answers:
1) _____ 2) _____ 3) _____ 4) _____ 5) _____ 6) _____

B) Name the triangle with the following properties.

7) This △ has an obtuse angle.   8) This △ has all acute angles.
9) This △ has no equal sides.   10) This △ has two equal sides.
11) This △ has a right angle.   12) This △ has three equal angles.
13) This △ has three equal sides.   14) This △ has equal base angles.

Answers:
7) _____ 8) _____
9) _____ 10) _____
11) _____ 12) _____
13) _____ 14) _____

C) Find the measure of an exterior angle in the following regular polygons.

15) Pentagon   16) Triangle   17) Octagon   18) Hexagon

Answers:

15) _____   16) _____   17) _____   18) _____

D) In each regular polygon, find the total number of interior degrees and the measure of an interior angle.

19) 10-sided figure   20) 5-sided figure

21) 6-sided figure   22) 4-sided figure

Answers:

	Total Interior Degrees	Interior Angle Measure
19)	_____	_____
20)	_____	_____
21)	_____	_____
22)	_____	_____

E) Find the missing measures.

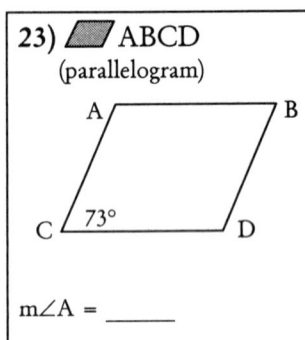

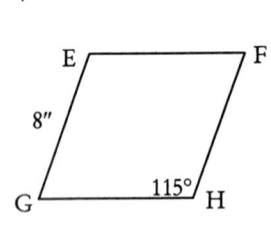

23) ABCD (parallelogram)

m∠A = _____
m∠B = _____
m∠D = _____

24) EFGH (rhombus)

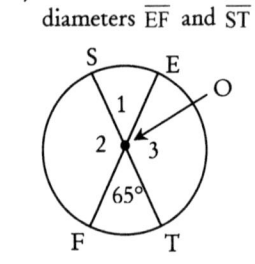

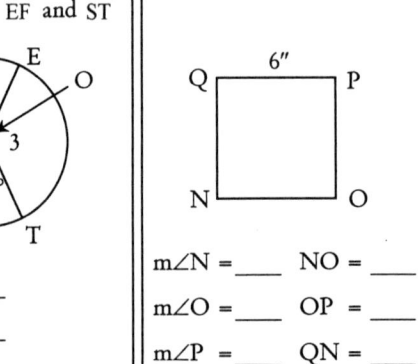

m∠E = _____   EF = _____
m∠F = _____   FH = _____
m∠G = _____   HG = _____

25) ⊙O with diameters $\overline{EF}$ and $\overline{ST}$

m∠1 = _____
m∠2 = _____
m∠3 = _____

26) NOPQ (square)

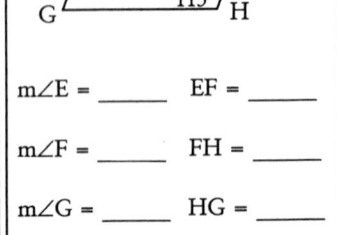

m∠N = _____   NO = _____
m∠O = _____   OP = _____
m∠P = _____   QN = _____
m∠Q = _____

Meeting the Challenge...Mathematics

F) Identify the following parallelograms as a rhombus, square, rectangle or none of these.

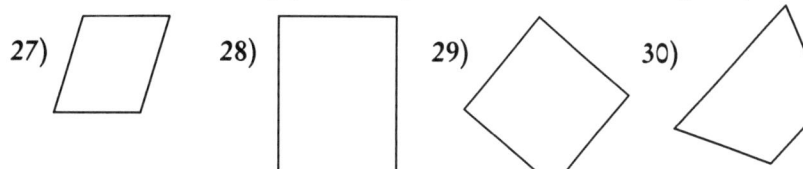

27)   28)   29)   30)

Answers:
27) _____   28) _____
29) _____   30) _____

G) The following objects have been rotated in a clockwise direction. Approximate the degree of rotation.

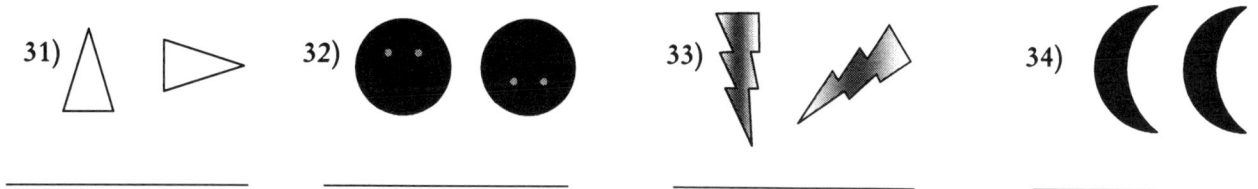

31)   32)   33)   34)

_____   _____   _____   _____

INDEPENDENT PRACTICE

A) Give the name of the polygon, the measure of an interior angle and the measure of an exterior angle.

	Name of Polygon	Total Interior Degrees	Interior ∠ Measure	Exterior ∠ Measure
1) 5-sided regular polygon				
2) 8-sided regular polygon				
3) 10-sided regular polygon				
4) 3-sided regular polygon				

B) Identify the following triangles by sides and by angles.

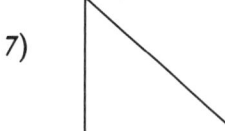

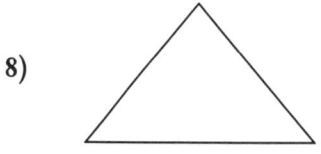

5)   6)   7)   8)

Sides	5)	6)	7)	8)
Angles				

C) List the special quadrilaterals (trapezoid, isosceles trapezoid, parallelogram, rectangle, rhombus and square) that are described below. There can be more than one quadrilateral in the answer.

9) both pairs of opposite sides parallel _____
10) parallelogram with right angles _____
11) quadrilateral with only one pair of opposite sides parallel _____
12) quadrilateral with opposite sides and opposite angles congruent _____
13) parallelogram with equal sides _____
14) quadrilateral with only one pair of opposite sides parallel and equal base angles _____

© 2003 Orange Frazer Press. All Rights Reserved.

Meeting the Challenge...Mathematics

D) A figure and its reflected image are given. Draw the line of reflection.

15)

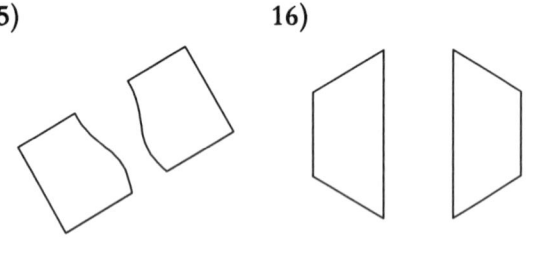

16)

E) Use the given line of reflection or symmetry to draw the reflection of the given figure.

17)

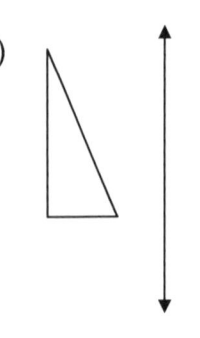

18)

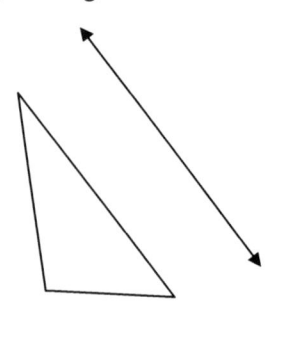

F) Look at each figure and give the most descriptive name.

19)

20)

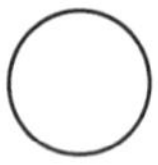

21)

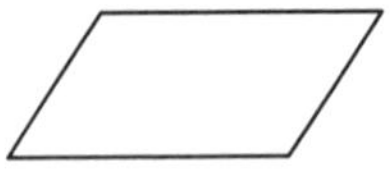

22)

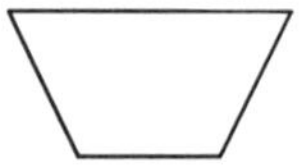

23)

24)

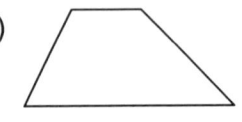

Answers:

19) _____ 20) _____

21) _____ 22) _____

23) _____ 24) _____

G) Use the appropriate letter to identify these three-dimensional objects.

25) 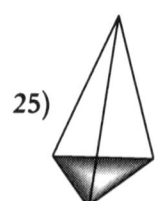   26) 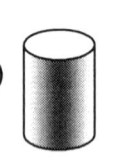   27)     28) 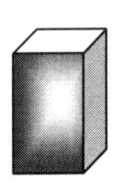   29) 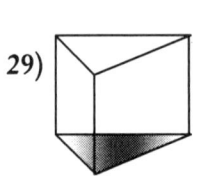   30)

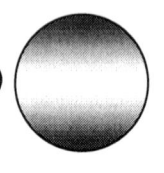

Answers:
25) ____
26) ____
27) ____
28) ____
29) ____
30) ____

a) cone   b) sphere   c) rectangular solid   d) cylinder   e) pyramid   f) triangular prism

Meeting the Challenge...Mathematics

H) Identify the following transformations as reflection, rotation or translation. Explain your reasoning by including specific information regarding the transformation. (i.e., draw the line of reflection, give the degrees of rotation and specify the direction, or give the direction of the translation)

31)

Answer: _____

Explanation:

32)

Answer: _____

Explanation:

I) Find the missing measures.

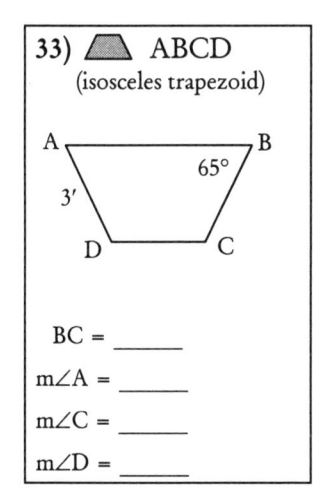

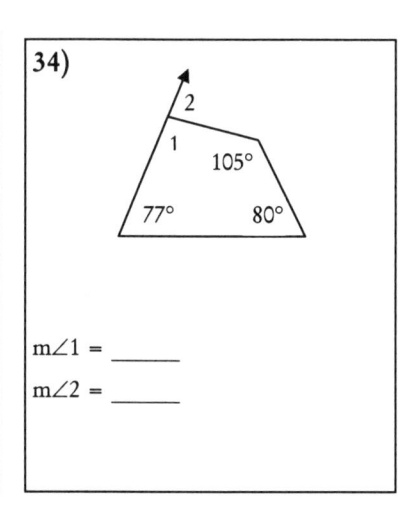

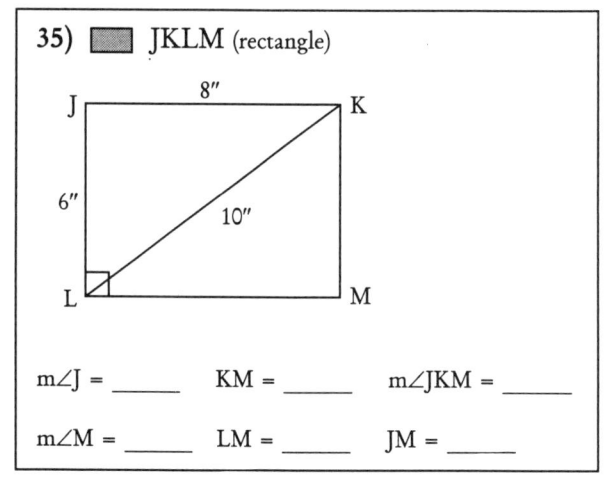

33) ABCD (isosceles trapezoid)

A — B, 65° at B, 3' at A side
D — C

BC = _____
m∠A = _____
m∠C = _____
m∠D = _____

34)

angles: 1, 2, 105°, 77°, 80°

m∠1 = _____
m∠2 = _____

35) JKLM (rectangle)

J — 8" — K
6"   10"
L   M

m∠J = _____   KM = _____   m∠JKM = _____
m∠M = _____   LM = _____   JM = _____

© 2003 Orange Frazer Press. All Rights Reserved.

J) Determine these values.

36) rhombus

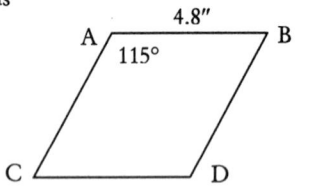

a) BD = ___ d) ∠B = ___

b) ∠C = ___ e) CD = ___

c) ∠D = ___ f) AC = ___

37) isosceles trapezoid

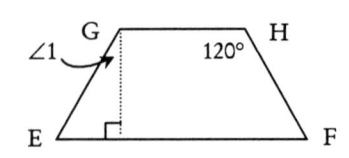

a) m∠1 = ___

b) ∠F = ___

c) ∠E = ___

38)

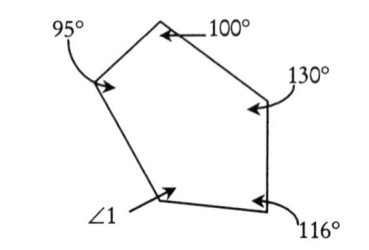

m∠1 = ___

39) $\overline{EF}$ is a diameter.

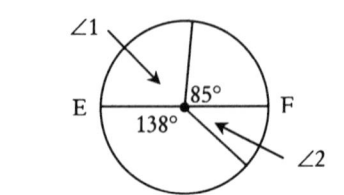

m∠1 = ___

m∠2 = ___

TEST PRACTICE

MULTIPLE-CHOICE QUESTIONS:

1) Give the name of a 6-sided regular polygon, the measure of an interior angle and the measure of ___
an exterior angle.

    a. pentagon, 90°, 60°   b. octagon, 136°, 44°   c. hexagon, 60°, 120°   d. hexagon, 120°, 60°

2) How does the measure of an interior angle of a regular polygon change as the number ___
of sides increases?

    a. increases   b. remains the same   c. decreases   d. approaches zero

Meeting the Challenge...Mathematics

3) The opposite pairs of angles in a parallelogram are always

    a. supplementary   b. complementary   c. vertical   d. equal

4) It is possible for a triangle to be

    a. both equilateral and obtuse   b. both scalene and isosceles

    c. both scalene and right   d. both isosceles and equiangular

5) The three-dimensional objects that have circular bases are

    a. pyramid and prism   b. cylinder and cone

    c. cube and pyramid   d. sphere and triangular prism

SHORT ANSWER QUESTIONS:

6) **Justify this statement:**
    If one angle of a parallelogram is a right angle, all other angles are also right angles.

7) If the sum of the measures of the interior angles of a polygon is 900°, how many sides does the polygon have?

The polygon has _____ sides.

8) Find the missing angle measures in the circle.

50°

m∠1 = _____

m∠2 = _____

m∠3 = _____

m∠4 = _____

Meeting the Challenge...Mathematics

EXTENDED RESPONSE QUESTIONS:

9) Jake needs to pack as many cans as possible in a rectangular box. The cans have a radius of 3 inches and are 8 inches in height. The box has dimensions of 14 inches × 12 inches × 18 inches. Jake may pack the cans in any way possible to maximize the number of cans packed. Make a diagram to demonstrate Jake's packing method.

The number of cans that Jake can pack is _____

10) Name each lettered figure with the word that best describes its appearance.

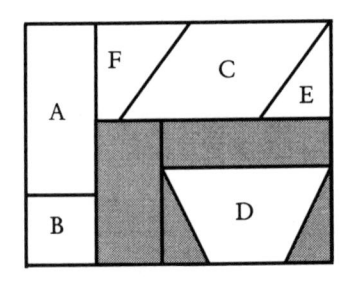

A) _____

B) _____

C) _____

D) _____

E) _____

F) _____

CLOSURE

1) Calvin is referring to a regular hexagon. All six sides are the same length and all six angles have the same measure.

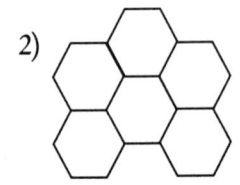

2)

3) Nature's prime example of the interlocking pattern of the hexagon is the bee's honeycomb.

90

Meeting the Challenge...Mathematics

MATHEMATICAL PROCESSES
Using Diagrams to Solve Problems

PURPOSE: In this lesson you will solve problems by developing a model or diagram, determining and obtaining the necessary information and deciding how to obtain it and verifying the solution.

A) A 6 meter ladder leaning against a building makes a 70° angle with the ground. The foot of the ladder is 3 ft. 6 in. from the base of the building. Tony plans to repair a broken window from this ladder.

1) Draw a sketch labeling the sides and dimensions and showing geometric relationships.

Sketch of the building and ladder

2) How far up the building does the ladder reach?

3) Tony does not like heights. If he is more than 10 feet above the ground he gets dizzy and can not do the job. Will Tony be able to climb the ladder high enough to fix the window?

B) The circle graph shows Mr. Hensley's budget for the family vacation to Lake Erie. He budgeted $850 for the vacation. The family spent $295 for food, $188 for gasoline, $250 for motels, $75 for recreation and $23 for miscellaneous items.

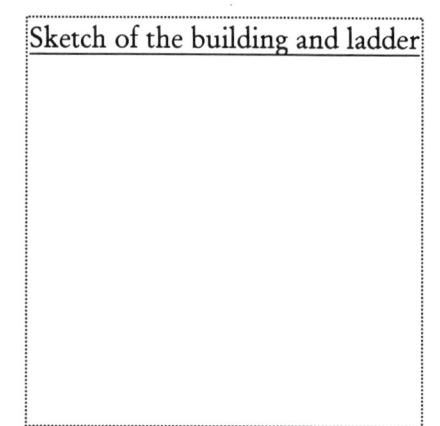

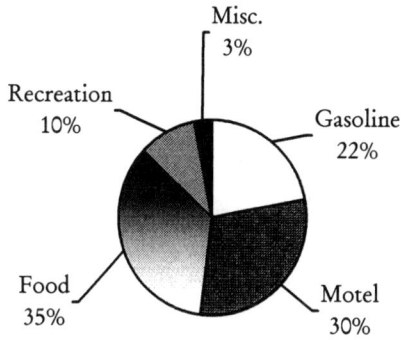

Misc. 3%
Recreation 10%
Gasoline 22%
Food 35%
Motel 30%

1) Discuss which items stayed within Mr. Hensley's budget and which items did not?

2) List some ways to help the family stay within each budget item in the future.

3) Did Mr. Hensley's family vacation stay within his budget?

C) The table lists gas mileage for a particular automobile at various speeds. As the speed of the automobile increases, so does the amount of gas that it uses.

Gasoline Mileage at Various Speeds

miles per hour	35	40	45	50	55	60
miles per gallon	32	29.5	27	25	23	24

1) Find the total number of gallons of gas that Shawna uses if she travels at

   a) 55 m.p.h. for 2 hours and 20 minutes _____

   b) 40 m.p.h. for 45 minutes _____

   c) 30 m.p.h. for one hour _____

2) What is the current price of a gallon of gasoline in your area? _____

3) If Shawna traveled two hours at 35 m.p.h, and two hours at 40 m.p.h., what is the difference in the number of gallons of gas used? _____

4) Use the table to estimate the miles per gallon if Shawna was traveling at:

   a) 65 m.p.h. _____   b) 70 m.p.h. _____

D) Shrubs are going to be planted along the perimeter of a plot of land in the shape of a right triangle with legs of 120 feet and 160 feet. The shrubs will be planted 5 feet apart with one shrub in each corner of the triangle.

1) Draw a sketch and show the dimensions of the plot of land and give an idea of how the shrubs will be distributed.

2) Exactly how many shrubs are needed? _____

3) The shrubs are priced at $14.95 each if you buy fewer than 10. If you buy 10 to 30 shrubs, the price is $13.95 each. If you buy more than 30 shrubs, the price is $10.95 each. Find the total cost of the shrubs. _____

Meeting the Challenge...Mathematics

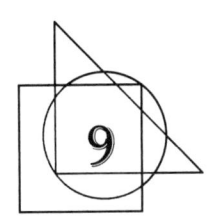

REPRESENTING MATHEMATICAL RELATIONSHIPS

Represent mathematical relationships in a variety of ways.

FOCUS: Mathematical relationships can be expressed in a variety of different forms. Let's look at an example.

The Adams family wants to save money by cleaning their own carpet. The cost to rent a carpet cleaning machine is $45 for the first day plus $25 for each additional day. Mr. Adams estimates that it will take them almost a week to complete the job. They checked with the Superior Carpet Cleaning Company and received a quote of $140. Should the Adams family clean the carpet themselves or have Superior do it?

1. Develop a rule or formula that relates the cost of the carpet machine rental (r) to the number of days (d) required to get the job done.

2. Use the rule or formula to complete a table that shows the cost of the machine rental for up to a week.

Number of days (d)	1	2	3	4	5	6	7
Rental cost (r)							

3. After how many days does it become cheaper for the Adams family to use the professional cleaner?

4. It took the Adams family 3 days to complete the cleaning job. Did they save money by renting the carpet cleaning machine? If so, how much?

PURPOSE: When you have completed this lesson, you will be able to represent a mathematical relationship algebraically from tables, charts, graphs, symbols or words. You will be able to translate among these various forms. You will also be able to describe how changing the value of one variable in an expression can have a direct effect on other variables in the expression.

WHAT YOU NEED TO KNOW

MATHEMATICS AND PROBLEM SOLVING VOCABULARY:

1. Constant of variation
2. Dependent variable
3. Direct variation
4. Horizontal axis
5. Independent variable
6. Linear variation
7. Origin
8. Vertical axis

© 2003 Orange Frazer Press. All Rights Reserved.

Meeting the Challenge...Mathematics

MEMORIZE THESE FACTS:

1. To identify a mathematical relationship, create a table from word sentences. Sometimes translating a table to a graph gives a better picture of the relationship. On a graph, the horizontal axis meets the vertical axis at the origin (zero point).

2. Direct variation is a relationship between two variables, x and y, of the form $y = mx$ where **m** represents the non-zero constant of variation. In this relationship the dependent variable (y) varies directly as the independent variable (x). It can also be said that y is directly proportional to x with **m** as the value of $\frac{y}{x}$ and referred to as the constant of proportionality. As y and x vary, their quotient, $\frac{y}{x}$, remains constant.

BE ABLE TO PERFORM THESE OPERATIONS:

Identifying Mathematical Relationships

Identify mathematical relationships from tables, graphs, symbols and words. Examine each pair of values and look for a mathematical operation or a combination of mathematical operations that takes the value of one variable (independent) to the value of the other variable (dependent).
Try this approach:

Step 1: Look for a simple one operation relationship such as addition, subtraction, multiplication, division, taking the square root or squaring. Look at the mathematical relationship when the independent variable (x in the table below) has a given value of zero. Look to the value of the corresponding dependent variable (y in the table below) for clues.

Examples:

a) Since $y = 5$ when $x = 0$, it suggests that there might be an addition relationship. Verify your idea by looking at other pairs of values from the table.

x	−2	−1	0	1	2	3
y	3	4	5	6	7	8

Mathematical relationship:
Add 5 to x to get y
Answer: $y = x + 5$

b) Since $y = 0$ when $x = 0$, it suggests that there might be a multiplication or division relationship. Verify your idea by looking at other pairs of values from the table.

x	−2	−1	0	1	2	3
y	−8	−4	0	4	8	12

Mathematical relationship:
Multiply x by 4 to get y
Answer: $y = 4x$

Step 2: Look for a multiple step relationship such as: a) multiply then add; b) divide then add, etc.
Example:

s	1	2	3	4	5
t	−1	1	3	5	7

Mathematical relationship: From the table we know that t increases by 2 when s is increased by 1. This means that t increases twice as fast as s and suggests an equation with $2s$ within it. **Answer:** $t = 2s - 3$

© 2003 Orange Frazer Press. All Rights Reserved.

Meeting the Challenge...Mathematics

Showing Mathematical Relationships

Graph a mathematical relationship, table or formula. Use this approach:
Step 1: Prepare a table of values.
  Example: $D = r \cdot t$ and $r = 60$ miles per hour

Time traveled in hours (t)	1	1.5	2	2.5	3	3.5	4
Distance traveled in miles (D)	60	90	120	150	180	210	240

Step 2: Draw and label the *x*-axis and the *y*-axis.
Step 3: Plot the points.
Step 4: Connect the points.

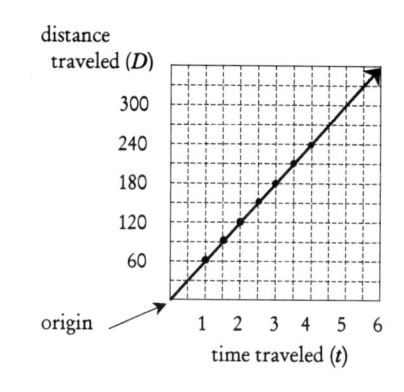

Using Mathematical Relationships

1. Find the missing values for points on a line graph. Look for a mathematical relationship between the variables, or set up a table to find the missing corresponding values.

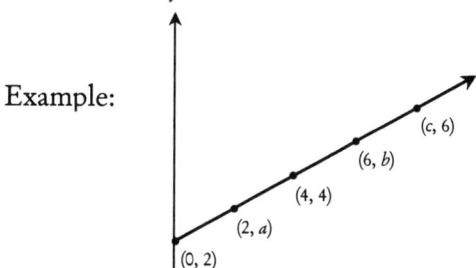

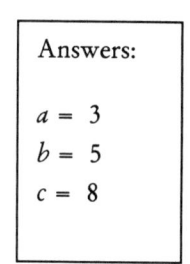

   Example: points $(0, 2)$, $(2, a)$, $(4, 4)$, $(6, b)$, $(c, 6)$

   Answers:
   $a = 3$
   $b = 5$
   $c = 8$

2. In an expression, identify how a change in the value of one variable (independent variable x) affects the value of another variable (dependent variable y).

   Example 1: Complete the table for $y = 5x$.

x	−2	−1	0	1	2	3
y	a	b	c	d	e	f

   $a = -10 \quad b = -5 \quad c = 0 \quad d = 5 \quad e = 10 \quad f = 15$

   a) As x increases, explain how y changes. **Answer:** As x increases by 1, y increases by 5.
   b) As x decreases, explain how y changes. **Answer:** As x decreases by 1, y decreases by 5.

   Example 2: In the formula $A = 6b$, b is the independent variable and A is the dependent variable. If b is positive and multiplied by 4, explain the effect on A. If $b = 9$, then $A = 54$. Now change the value of b by multiplying by 4 so that $b = 36$ and $A = 216$. **Answer:** If b is multiplied by 4, so is A.

   Example 3: In the formula $a = 30 \div b$, if b is positive and cut in half, tell what happens to a. **Answer:** If $b = 10$, then $a = 3$. If b is positive and cut in half, $b = 5$ and $a = 6$. Therefore, the value of a will be positive and will double in value as a result of the change in b.

BE ABLE TO PERFORM THESE CALCULATOR OPERATIONS:

1. Use the positive/negative key $[+/-]$. Enter a number. Use this key to change the sign (take the opposite).
2. Use the division key $[\div]$ to find the constant of variation of a ratio.
3. Use the square root key $[\sqrt{x}]$. Enter a number and use this key to find its square root.
4. Use the square key $[x^2]$. Enter a number and use this key to square it.

GUIDED PRACTICE

A) Complete the table and give a formula to calculate the corresponding values.

1)

Number of gallons (g)	1	2	3	4	5	6
Cost (c)	$1.15	$2.30	$3.45	a	b	c

$a =$ _____ $b =$ _____ $c =$ _____ Formula: _____

2)

f	1	2	3	4	5	6	7	8	9
g	4	7	10	13	m	n	22	r	s

$m =$ _____ $n =$ _____ $r =$ _____ $s =$ _____ Formula: _____

3)

j	1	2	3	4	5	6	7	d	9	10	f	12
k	1	4	a	b	25	c	49	64	e	100	121	g

$a =$ _____ $b =$ _____ $c =$ _____ $d =$ _____ $e =$ _____ $f =$ _____ $g =$ _____

Formula: _____

B) Complete the following tables using the given mathematical relationships.

4) Kendra is six years older than Jackie.

Jackie's age in years (J)	5	6	15	16	25	35
Kendra's age in years (K)	a	b	c	d	e	f

$a =$ _____ $b =$ _____ $c =$ _____ $d =$ _____ $e =$ _____ $f =$ _____

Formula: _____

Meeting the Challenge...Mathematics

5) The rate of a car rolling down a hill in neutral varies directly to the length of time it rolls. Fill in the table.

Rate in feet per second	18	x	y	z
Time in seconds	6	10	15	20

x = _____ y = _____ z = _____

6) A formula used to relate a man's foot length in inches (l) to shoe size (s) is $s = 3l - 25$. Use this formula to fill in the table below. Shoes come in whole and half sizes. Graph your answers.

Foot length in inches (l)	10	10.5	11	11.5
Shoe size (s)	a	b	c	d

a = _____ b = _____ c = _____ d = _____

shoe size

Draw, label and describe the graph. _____

If a man wears a size 7 shoe, about how long is his foot? _____

If the length of a man's foot is 12 inches, what shoe size would he wear? _____

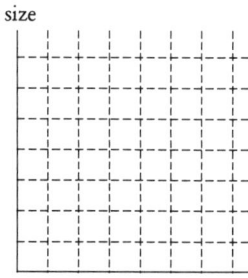

foot length

C) Find the missing coordinate values on the line graph.

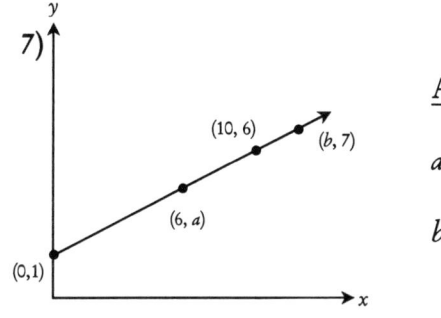

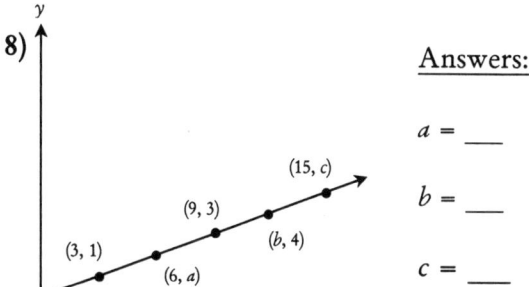

7) Points on graph: (0,1), (6, a), (10, 6), (b, 7)

Answers:
a = ___
b = ___

8) Points on graph: (3, 1), (6, a), (9, 3), (b, 4), (15, c)

Answers:
a = ___
b = ___
c = ___

D) Explain how the change in one variable affects the other.

9) Formula: $A = s^2$

If s is doubled, tell how A is affected.

_____

10) Formula: $B = \dfrac{25}{a}$

If a is positive and increases, tell how B is affected.

_____

INDEPENDENT PRACTICE

A) Give the formula that relates *x* and *y*.

1)

x	0	1	2	3	5
y	2	5	8	11	17

Formula: _____

2)

x	0	2	4	7	10
y	−3	7	17	32	47

Formula: _____

3)

x	0	2	4	6	8	10
y	1	2	3	4	5	6

Formula: _____

4)

x	36	25	16	9	4	1
y	6	5	4	3	2	1

Formula: _____

B) Find the missing values in the table and state the formula used to determine those values.

5)

a	−2	−1	0	1	2	3	4	m
b	8	j	0	−4	k	−12	l	−20

j = ____   k = ____   l = ____   m = ____   Formula: _____

6)

a	3	−1	4	7	8	y	−3	d
b	36	c	48	84	x	120	−36	0

x = ____   y = ____   c = ____   d = ____   Formula: _____

C) Evaluate each formula using the given numbers.

7) $m = 6y$

m	−18	−6	0	d	9	f
y	a	b	c	0.5	e	2

a = ____   b = ____   c = ____   d = ____   e = ____   f = ____

Meeting the Challenge...Mathematics

8) $b = 2a - 5$

a	x	−3	0	c	−1	e
b	7	y	z	−3	d	0

x = ____   y = ____   z = ____   c = ____   d = ____   e = ____

D) Use the graph and mathematical relationships to find the missing values.

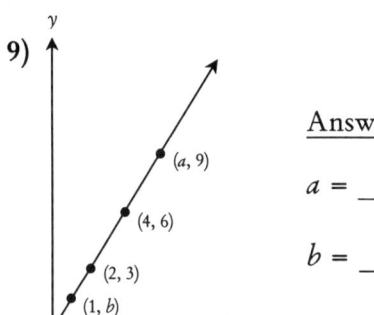

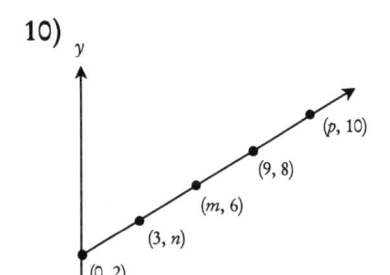

9) Points on graph: (1, b), (2, 3), (4, 6), (a, 9)

Answers:
a = ____
b = ____

10) Points on graph: (0, 2), (3, n), (m, 6), (9, 8), (p, 10)

Answers:
m = ____
n = ____
p = ____

E) Give a formula in terms of C and D to calculate the values in column E.

11) Formula: E = _____

C	D	E
1	3	10
2	5	17
3	4	15
4	2	10

12) Formula: E = _____

C	D	E
2	0	2
3	1	1
5	1	3
6	2	2

F) Answer the questions regarding the mathematical relationships.

13) Each week Jason is paid $75 plus a commission of 5% of his sales. Fill in the table and write an equation that relates Jason's weekly income(I) and his sales (s). Label the axes and plot the points on the graph.

Sales in dollars (s)	0	500	1000	1500	2000
Sales commission					
Weekly income (I)					

Equation: _____

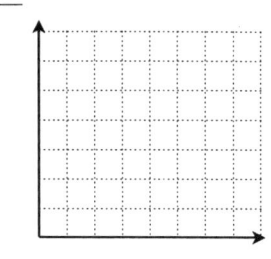

If Jason had $2700 in sales, plot the point on your graph that represents his income.

Amount of income: _____

14) The Ridge Avenue Parking Garage charges a flat rate of $1.50 an hour. The South Street Parking Garage nearby charges $2.00 for the first hour and $1.25 for each additional hour.

   a) State a formula to find the cost of parking at each garage for *h* hours.

   Ridge Avenue Parking Garage: _____

   South Street Parking Garage: _____

   b) Compare the cost of the two garages by completing the table below.

Hours parked	1	2	3	4	5	6
Cost at Ridge Avenue						
Cost at South Street						

   c) The cost for parking is the same at either garage if a person parks for _____ hours.

   d) Which garage has the cheaper parking for 2 hours? _____

   e) Which garage has the cheaper parking for 5 hours? _____

TEST PRACTICE

MULTIPLE-CHOICE QUESTIONS: Write the letter of the correct answer.

1) Study the table and then choose the value to complete this formula: $y = 2x + ?$

x	1	2	3	4	5
y	3	5	7	9	11

   a. 0   b. 1   c. −1   d. 2

2) The following table shows the mathematical relationship of the area and height of several parallelograms. Find the value of the base.

Parallelogram	a	b	c	d	e	f
Height	4	5	8	9	10	12
Area	24	30	48	54	60	72

   a. 12   b. 8   c. 6   d. 4

3) The minimum charge for using city water is $25 per month for up to 5000 gallons. For each gallon (*g*) used over 5000, there is an additional charge of $0.08 per gallon. Tell which expression calculates the amount (*A*) of the water bill when monthly usage goes over 5000 gallons.

   a. $A = 25 + 5000g$ b. $A = (5000)(0.08)g$ c. $A = 25 + 0.08(g - 5000)$ d. $A = 25 + 0.08g$

Meeting the Challenge...Mathematics

4) Brian is paid $3 an hour to mow lawns plus a $5 flat fee for the use of his lawnmower. Choose the equation that can be used to determine Brian's earnings (E) for hours (H) worked.

   a. $E = 8H$ b. $E = 5(3 + H)$ c. $E = 5 + 3H$ d. $E = 5H + 3$

5) Look at the graph. Choose the formula used to determine the points on the line.

   a. $y = \frac{1}{3}x$ b. $y = 2x$

   c. $y = 3x$ d. $y = \frac{1}{2}x$

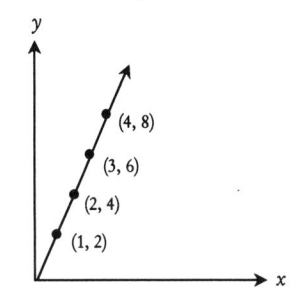

   (4, 8)
   (3, 6)
   (2, 4)
   (1, 2)

SHORT ANSWER QUESTIONS:

6) Consider this formula: $x = ab$
   a) If a is doubled and b is cut in half, explain how the value of x changes.

   b) Give an example to justify your explanation.

   Explanation: _____

   Example: _____

7) Find the missing values in the table and give a formula that relates x and y.

x	−2	−1	0	1	2	b
y	−10	−5	0	a	10	15

Answers:

$a = $ ____ $b = $ ____

Formula: _____

8) x varies directly as $y + 3$.

   a) Write an expression for this relationship.

   _____

   b) If $y = 5$ when $x = 2$, find y when $x = 6$.

   Show your work to find y. $y = $ ____

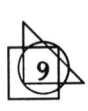

EXTENDED RESPONSE QUESTIONS:

9) Consider this formula: $y = 10 - 2x$

   a) If x is positive and increases, explain how y is affected.
   _____

   b) If x is negative and increases, explain how y is affected.
   _____

10) To make a long distance phone call, Gail could use a calling card that charges a $0.49 connection fee plus $0.03 a minute, or her regular long distance service at $0.10 a minute.

   a) State the formula to find the cost of the call using the calling card. Formula: _____

   b) State the formula to find the cost of the call using the long distance service. Formula: _____

   c) Which method is cheaper for a 10 minute call? _____

   Why? _____

CLOSURE

The main objective for the Adams family is to save money on their carpet cleaning. Developing and using mathematical relationships can help to compare costs between renting a machine and hiring a professional.

1) Translate the mathematical relationship for renting the carpet machine into a formula.
   Formula: $r = 45 + 25(d - 1)$

2) Table of rental cost

Number of days (d)	1	2	3	4	5	6	7
Rental cost (r)	$45	$70	$95	$120	$145	$170	$195

3) Superior Carpet quoted $140 to do the job. If it takes the Adams family 5 days or more to do the job, it would be cheaper for them to use the professional company.

4) The Adams family saved $45 ($140–$95) by cleaning the carpet themselves in 3 days.
   It was a lot of work but the Adams family will have an extra $45 to spend on their family vacation.

Meeting the Challenge...Mathematics

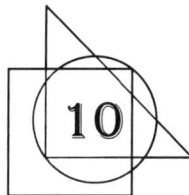

ALGEBRAIC EXPRESSIONS AND FORMULAS

Write, simplify, evaluate, and use algebraic expressions and formulas.

FOCUS: We may not use x and y everyday, but let's see how algebraic expressions and formulas can be used in our daily lives.

Traditionally, your high school class sells fruit as a money making project during the holidays. The class buys the fruit at a cost of $140.00 for the first 125 pounds. The cost for every pound over 125 pounds is $0.55 per pound.
1. Develop a formula to find the cost of the fruit knowing the amount of fruit ordered is more than 125 pounds.
2. Last December the class took orders for 565 pounds of fruit. Solve for the cost of the fruit.
3. Fruit sales totaled $638.00. Find the profit.

PURPOSE: When you have completed this lesson, you will be able to simplify, evaluate, construct, apply and interpret the meaning of algebraic expressions and formulas. Special mention will be given to understanding and using *Order of Operations* and the *Laws of Exponents*.

WHAT YOU NEED TO KNOW

MATHEMATICS AND PROBLEM SOLVING VOCABULARY:

1. Algebraic expression
2. Algebraic term
3. Degree of polynomial
4. Distributive property
5. Equivalent expressions
6. Evaluate
7. Factor
8. Grouping symbols
    (parentheses, brace, bracket, fraction bar)
9. Monomials
    a. Coefficient of a monomial
    b. Base and exponent of a monomial
10. Numerical coefficient
11. Opposite
12. Polynomial
13. Reciprocal
14. Similar or like terms
15. Variable

MEMORIZE THESE FACTS AND FORMULAS:

1. Order of Operations:
    Evaluate algebraic expressions by following the steps below:
    Step 1: Simplify expressions contained within grouping symbols (parenthesis, bracket, brace, fraction bar). If a pair of grouping symbols is contained within another, work from the inside or the innermost set outward.
    Step 2: Simplify terms with exponents.
    Step 3: Do all multiplication and division in order from left to right.
    Step 4: Do all addition and subtraction in order from left to right.

© 2003 Orange Frazer Press. All Rights Reserved.

2. Rules for performing math operations with positive and negative numbers:
   a. Addition
      - If both numbers are positive, add the numbers and the answer is positive.
      - If both numbers are negative, add the numbers and the answer is negative.
      - If one is positive and the other negative, subtract the numbers and take the sign of the number with the larger absolute value.
   b. Subtraction
      - Take the opposite of the number to be subtracted.
      - Add both numbers and apply the rules for adding positive and negative numbers.
   c. Multiplication
      - If both numbers have the same sign (two positives or two negatives), the answer is positive.
      - If one number is positive and the other is negative, the answer is negative.
   d. Division
      - If both numbers have the same sign (two positives or two negatives), the answer is positive.
      - If one number is positive and the other is negative, the answer is negative.
3. Key words and phrases for the following:
   - Addition - more than, increased by, added to, in addition to, plus, sum, deposit, rose, gain, above
   - Subtraction - less than, fewer than, decreased by, minus, take away, taken from, difference, dropped, loss, withdrawal, below
   - Multiplication - product, times, the word "of", twice
   - Division - quotient, goes into, cut into parts, divisible by, fraction of
   - Equation – is, equals, is the same as
4. Special case algebraic word problems:
   - Consecutive integers: Start with b, add one to find the next consecutive integer, $b + 1$, $b + 2$, etc.
   - Consecutive even and odd integers: If a is an even (odd) integer, add 2 to find the next consecutive even (odd) integer: $a + 2$, $a + 4$, $a + 6$, etc.
   - Money value: Multiply the value of the coin or bill by the number of coins or bills.
   - Distance problems: Pay special attention to the units used to express distance, rate and time. For example, if the rate is given in miles per hour, then time must be given in hours.
5. Laws of Exponents:
   - Multiplication – When multiplying monomials, add the exponents of terms with the same base.
   - Division – When dividing monomials, subtract the exponents of terms with the same base.

BE ABLE TO PERFORM THESE OPERATIONS:

Simplifying Expressions

1. **Apply the rules for working with positive and negative numbers.**
   - Addition examples: a) $-5 + (-3) + (-2)$ **Answer:** -10 b) $5 + (-17) + 6$ **Answer:** -6
   - Subtraction examples: a) $7 - 25$ **Answer:** -18 b) $-8 - 6$ **Answer:** -14
   - Multiplication examples: a) $4 \times (-3)$ **Answer:** -12 b) $-5 \times (-6)$ **Answer:** 30
   - Division examples: a) $-28 \div 4$ **Answer:** -7 b) $-32 \div (-8)$ **Answer:** 4

Meeting the Challenge...Mathematics

2. **Find and use the opposite of a number.**
   - Find the opposite of a number by changing its sign.
     Examples: a) The opposite of -3 is $+3$. b) The opposite of 6 is -6.
   - To subtract a positive or negative number, take its opposite and add.
     Examples: a) $-12 - (+3) = -12 + (-3)$ **Answer:** -15
     b) $9 - (-4) = 9 + 4$ **Answer:** 13

3. **Combine like terms.** Identify terms that contain the same variable to the same powers. Add or subtract the coefficients.
   Examples: a) $-3x^5 + 7x^5$ **Answer:** $4x^5$
   b) $5x - 7x$ **Answer:** $-2x$
   c) $12x - 5y - 6y - 3x + y$ **Answer:** $9x - 10y$

4. **Use the distributive property to remove parentheses.** Multiply the expression outside the parentheses by each term inside.
   Examples: a) $7x(5x + 2y)$ **Answer:** $35x^2 + 14xy$
   b) $2x^5(-3x^3 + 7x^2 - 4x)$ **Answer:** $-6x^8 + 14x^7 - 8x^6$
   c) $5a^3b(3ab - 2a^2 + b)$ **Answer:** $15a^4b^2 - 10a^5b + 5a^3b^2$

5. **Multiply or divide monomials using the Laws of Exponents.**
   Step 1: Multiply or divide the numerical coefficients. Pay attention to the positive and negative signs.
   Step 2: When multiplying monomials, add the exponents of factors with the same base.
   Step 3: When dividing monomials, subtract the exponents of factors with the same base.

   Examples: a) $(5d^3e^2f)(-3d^2ef^7)$ **Answer:** $-15d^5e^3f^8$

   b) $-\dfrac{15x^5}{18x^2}$ **Answer:** $-\dfrac{5x^3}{6}$

Evaluating Expressions

1. **Evaluate arithmetic expressions.** Use Order of Operations.
   Step 1: Simplify grouped expressions and simplify terms with exponents.
   Step 2: Multiply and divide in order from left to right; add and subtract in order from left to right.
   Examples: a) $4 + (-5)(-9) + (-16) = 4 + 45 + (-16)$ **Answer:** 33
   b) $3[14 + (-9)] - 6 = 3(5) - 6 = 15 - 6$ **Answer:** 9
   c) $\{3(2 + 6)\}(5) = \{3(8)\}(5) = \{24\}(5)$ **Answer:** 120
   d) $5(-4)^3 = 5(-4)(-4)(-4)$ **Answer:** -320

2. **Evaluate algebraic expressions using order of operations.**
   Step 1: Replace each variable with the given value.
   Step 2: Simplify grouped expressions and simplify terms with exponents.
   Step 3: Multiply and divide in order from left to right; add and subtract in order from left to right.

   Examples: a) $5d - 2g$, where $d = 3$, $g = -5$ | b) $\dfrac{v}{s+w}$, where $s = 1$, $v = 6$, and $w = 0.5$

   $5(3) - 2(-5) = 15 + 10$ **Answer:** 25 | $\dfrac{6}{1+0.5} = \dfrac{6}{1.5}$ **Answer:** 4

3. Evaluate monomials using the Laws of Exponents.

   Examples: a) $2a^3$, $a = 5$ Substitute: $2 \times 5^3 = 2 \times 5 \times 5 \times 5$ **Answer:** 250

   b) $7c^2d$, $c = -3$, $d = -2$ Substitute: $7(-3)^2(-2) = 7(-3)(-3)(-2)$ **Answer:** -126

   c) $6a^3 - 2ab$, $a = 3$, $b = -2$ Substitute: $6(3)^3 - 2(3)(-2) = 6 \cdot 27 + 12$ **Answer:** 174

Writing and Interpreting

1. **Determine operations from key words and phrases.**

   Examples: a) Word phrase: 5 decreased by twice a number
   Mathematics phrase: $5 - (2 \times \text{a variable})$

   b) Word phrase: 12 more than half a number
   Mathematics phrase: $12 + (\frac{1}{2} \times \text{a variable})$

2. **Convert word phrases to algebraic expressions.**

   Examples: a) Word phrase: d less than 3 times n Algebraic expression: $3n - d$

   b) Word phrase: 12 more than $\frac{2}{3}$ of b Algebraic expression: $12 + \frac{2}{3}b$

3. **Write an algebraic equation from words.**

   Step 1: Look for a mathematical relationship given by a word phrase or sentence.
   Step 2: Replace word phrases with algebraic symbols and expressions.
   Step 3: Relate the algebraic symbols and expressions in the equation.

   Example: Three greater than twice another number is 42.
         3   +   2 • b = 42

   Equation: $3 + 2b = 42$

4. **Construct algebraic expressions from special case algebraic problems.**
   - Represent the relationship of two numbers.
     Examples: a) One number is 6 times another.
          Algebraic representation: One number = c, the other = $6c$
     b) The length of a rectangle is 4 times its width.
          Algebraic representation: width = w, length = $4w$
     c) The sum of two numbers is 25.
          Algebraic representation: One number = x, the other = $25 - x$
   - Represent the value of money (coin problems).
     Example: Represent the value of d quarters.
          Algebraic representation: value = $0.25d$
   - Represent the distance traveled using $d = r \times t$, paying careful attention to units.
     Example: Represent the distance traveled in 45 minutes if a vehicle travels at a rate of 55 miles per hour.
          Algebraic representation: $d = 55 \times \frac{3}{4}$ (or 0.75) Convert 45 minutes to ¾ hour.

BE ABLE TO PERFORM THESE CALCULATOR OPERATIONS:

1. Use the positive/negative key $\boxed{+/-}$. Enter a number. Use this key to change the sign (take the opposite).
2. Use the square key $\boxed{x^2}$ to evaluate monomials when possible. Enter the value for x, then press the $\boxed{x^2}$ key.

Meeting the Challenge...Mathematics

GUIDED PRACTICE

A) Simplify.

1) $-9 + 6 + 7 + (-8) =$ _____

2) $-[-125 + 200] - [96 - 150] =$ _____

3) $(-6)(4) - \left(\dfrac{-21}{3}\right) =$ _____

B) Combine like terms.

4) $6x - 3y - 5x + 8y + 2x - 7y =$ _____

5) $-2(x - 4y) + 3(x + y) =$ _____

C) Simplify.

6) $2(5x - 3) - 4(x - 2) =$ _____

7) $3x - (5 - 2x) + 8x - 15 =$ _____

D) Simplify.

8) $(2a^3)(-5a^2) =$ _____

9) $(4x^5)^3 =$ _____

10) $(-3x^2)(2x^3)^2 =$ _____

11) $\dfrac{48b^3c^2}{6bc^5} =$ _____

12) $\dfrac{(2x^4)^3}{(-3x^2)^5} =$ _____ = _____

E) Evaluate for $x = 3$, $y = -8$, $z = -2$

13) $x + y - z =$ _____

14) $3y^2 =$ _____

15) $-5(2y - 3z) =$ _____

16) $2yz + xz =$ _____

17) $-5x^2yz^2 =$ _____

107

F) **Evaluate the formulas.**

18) The distance traveled during acceleration is calculated using the formula, $D = \dfrac{at^2}{2}$.

    If $a = 13.7$ and $t = 25$, solve for D.

    $D =$ _____

19) The formula for the area of a trapezoid is $A = \dfrac{h(b+B)}{2}$.

    If $h = 4$ inches, $b = 6$ inches, and $B = 7$ inches, solve for A.

    $A =$ _____

G) **Express as algebraic phrases.**

20) a number decreased by 12 _____

21) y less than 15 _____

22) twice a number increased by 8 _____

23) 25 cut into d parts _____

H) **Create a formula for the following rules.**

24) Change f feet to i inches. _____

25) Total cost (c) of n articles at d dollars each _____

I) **Create a formula using the following geometric information.**

26) The three sides of a triangle are b, $b - 3$, and $b + 4$. Write the formula for the perimeter. Simplify.

    $P =$ _____ $=$ _____

27) In a rectangle the length (l) is 12 more than the width (w).

    Express the length and width in algebraic terms.

    $l =$ _____ $w =$ _____

    Express the perimeter in algebraic terms. Simplify.

    $P =$ _____ $=$ _____

Meeting the Challenge...Mathematics

J) Develop an equation from the word phrases and sentences.

28) If one-third of a number is diminished by 16, the result is 21.

Equation: _____

29) Jordan weighs twice as much as his brother. The sum of their weights is 375 pounds.

Represent algebraically:

Jordan's weight: _____   brother's weight: _____

Equation: _____

INDEPENDENT PRACTICE

A) Simplify.

1) $-5 + (-2)(4) + (2)(9) = $ _____

2) $28(-5) + 4 \cdot 2 + (-5)\,2 = $ _____

3) $\left(\dfrac{-3}{4}\right)\left(\dfrac{1}{2}\right)\left(\dfrac{-2}{9}\right) = $ _____

4) $8(-3) + (-1)(-6) - (-2)(4) = $ _____

5) $\dfrac{4(5+1)}{8(15-13)-4} = $ _____

6) $(-4)^2 = $ _____

7) $[5^2 - (3)(-2)(-3)]^2 = $ _____

8) $50 + (-2)[\,6 + (-3)^3(4)] = $ _____

9) $\left(\dfrac{-2}{3}\right)^2 = $ _____

10) $6(-5) + (2)(7) - (-2)(-4) = $ _____

11) $\left[\left(\dfrac{3}{5}\right)-\left(\dfrac{-7}{10}+\dfrac{2}{5}\right)\right] =$ _____

12) $(6)^3(-2)^4 =$ _____

13) $[4^2+(2)(6)(-3)]^2 + 2(-2)^3 =$ _____

14) $[30 \div (5 \times 2)] \div (-3) =$ _____

15) $\dfrac{2(-1)(-5)-(6)(5)}{2(7)-9} =$ _____

16) $27 \div (-9) + [(27+9) \div 3] =$ _____

17) $3 + 48 - 16 - 35 \div 7 =$ _____

18) $\dfrac{9-5(3-2)}{3+5} =$ _____

19) $\dfrac{(6-2)(6+2)^2}{4} + (6 \bullet 3) - 2 =$ _____

20) $8^2 - [2 \times 3 - (4+1)^2] =$ _____

Meeting the Challenge...Mathematics

B) Evaluate for $a = \frac{1}{2}$, $b = \frac{2}{3}$, $c = \frac{1}{4}$

21) $6abc =$ _____

22) $4a - 8c =$ _____

23) $2b^3 =$ _____

24) $(a + b)^2 =$ _____

25) $abc(3a - 4bc) =$ _____

26) $\frac{a}{c} + 6b =$ _____

27) $c^2 + b^2 + a^2 =$ _____

C) Evaluate the formulas. Show the substitutions and your work.

28) The formula to convert degrees Fahrenheit to degrees Centigrade is $C = \frac{5}{9}(F - 32)$.

If $F = 99°$, solve for C.

C = _____

29) The formula to solve for electrical power is $P = I^2 R$.

If $I = 15$ amps and $R = 2.5$ ohms, solve for P.

P = _____

30) One formula that can be used to find the normal weight of a person who is at least 5 feet tall is $w = \frac{11(h-40)}{2}$. The normal weight in pounds is represented by w, and h represents the person's height in inches. If Cassie is 5 ft. 4 in. tall, calculate her normal weight.

$w = $ _____

31) The formula to find the slope of a line is $\frac{Y-y}{X-x}$.

Let $Y = 7$, $y = 3$, $X = 6$, and $x = 2$.

Slope = _____

D) Use $a = 4$, $b = -5$, $c = 3$, and $d = -3$ to evaluate these expressions.

32) $a(b + c) + d = $ _____

33) $-a(c - d) = $ _____

34) $4b^3 = $ _____

35) $(ad)^2 = $ _____

36) $(b + c)^2 = $ _____

37) $(b - c)^2 = $ _____

38) $ab + cd = $ _____

39) $-5a^2 b^3 c = $ _____

E) Combine like terms.

40) $5a - 4b + 3a + b = $ _____

41) $-2x + 5y - 3x + 4z - x + y - 3z = $ _____

42) $6x - 7y + 12x + 9y = $ _____

43) $3x^2 - 5y^2 - x^2 + y^2 = $ _____

44) $12x + 19 + (-8x) + 5 = $ _____

45) $-12d^2 - (-2d^2) - 6d^2 = $ _____

46) $5x^2 + 3x - 2x^2 - 8x + 3x^2 = $ _____

47) $4ab + 2bc - 2ab + 3bc = $ _____

Meeting the Challenge...Mathematics

F) Remove the parentheses and combine like terms.

48) $9a + 4(2a) + (-6)(3a) - a$

49) $10y^2 + 4 - (3y^2 - 1)$

50) $4x + 3xy - (9x - 4xy)$

51) $2(15a - 6b) - (-2a + 3b - 5)$

52) $3x - 3(4 - 2x) + 5$

53) $4(y - 3) - 5(2y + 1)$

54) $3a^2 + 2(4a^2 - b^2) + 10b^2$

55) $4c - (2c^2 + 3c + 6) + 2(4c^2 - 5)$

G) Translate word phrases to algebraic expressions and equations.

56) A wheat roll has 10 fewer calories than twice the number of calories in a slice of white bread. Use algebra to represent the calories contained in each.

Wheat roll = _____

Slice of white bread = _____

57) Tamara owns x cds. Nicole owns 4 more than twice as many as Tamara.

a) Give an algebraic expression that represents the number of cds owned by Nicole.

Number of Nicole's cds = _____

b) Give an algebraic expression that represents the number of cds that Tamara and Nicole own together.

Total cds = _____

58) Each of the stripes on the American flag is *w* inches wide. Give the algebraic expression that represents the total width of the stripes.

Total width of the stripes = _____

59) Seven sections of paneling are needed to panel one wall in the family room. Each section is *p* inches wide. Express the length of the wall algebraically.

Length of the wall (in inches) = _____

Length of the wall (in feet) = _____

60) Rico is the youngest child in the family. His brother Joe is twice as old as Rico, and his sister Monica is 4 years younger than Joe. Represent their ages algebraically.

Rico's age = _____

Joe's age = _____

Monica's age = _____

61) An oak tree is half as tall as an elm tree. Let *h* be the height of the elm tree. Represent the height of the oak tree.

Height of the oak tree = _____

62) Represent algebraically the money value of the following coins and bills.

    a) *q* quarters = _____   b) *d* dimes = _____   c) *n* nickels = _____   d) *p* pennies = _____

    e) *w* one-dollar bills = _____   f) *c* five-dollar bills = _____   g) *h* ten-dollar bills = _____

63) Represent algebraically.

    a) Four consecutive even integers   1st _____   2nd _____   3rd _____   4th _____

    b) The sum of four consecutive even integers is 60. Equation: _____

    c) Two integers whose sum is 90   _____ , _____

Meeting the Challenge...Mathematics

TEST PRACTICE

MULTIPLE-CHOICE QUESTIONS: Write the letter of the correct answer.

1) $(5a)^4$
   a. $5 \cdot a \cdot a \cdot a \cdot a$
   b. $5a \cdot 5a \cdot 5a \cdot 5a$
   c. $5 \cdot a \cdot 4$
   d. $5a + 5a + 5a + 5a$

2) $4x + 8y - 3x - 5y + 7$
   a. $4xy + 7$
   b. $7x + 3y + 7$
   c. $x + 3y + 7$
   d. $12x - 8y + 7$

3) The product of $5x^4y$ and $-4x^3y^3$ is
   a. $-20x^{12}y^4$
   b. x^7y^4
   c. $-20x^{12}y^3$
   d. $-20x^7y^4$

4) For $a = -2$ and $b = 3$, the value of $6a + b^2$ is
   a. 21
   b. -3
   c. 22
   d. -6

5) $6x + 13 - 4(2x - 4)$
   a. $-2x - 3$
   b. $-2x + 29$
   c. $14x + 29$
   d. $2x - 3$

SHORT ANSWER QUESTIONS:

6) Leslie's bowling score was 25 points more than half of Ben's score.

   a) Represent Leslie's and Ben's bowling scores algebraically. Leslie: _____ Ben: _____

   b) Leslie bowled 138. Set up an equation to solve for Ben's score. Equation: _____

7) A plot of land with an area of 65,000 square feet is subdivided into four building lots. Lots 2 and 3 each have twice the area of Lot 1. Lot 4 is 5,000 square feet larger in area than Lot 1.

   a) Represent the size of each lot algebraically.
   Lot 1: _____   Lot 2: _____   Lot 3: _____   Lot 4: _____

   b) Set up an equation for this word problem.
   Equation: _____

8) Fifteen loaves of bread are made from a batch of dough. The total weight (x) of the batch of dough is given in pounds.

   a) Represent algebraically the weight of each loaf. _____

   b) Add three pounds of dough to the batch. Represent algebraically the weight of each loaf. _____

   c) Double the batch in part b. Represent algebraically the weight of each loaf. _____

EXTENDED RESPONSE QUESTIONS:

9) To have fliers copied one must pay $8.00 for the first one hundred copies and $5.00 for each additional hundred.

   a) Explain how to figure the price for more than one hundred copies.
   _____

   b) Write an equation for the total cost (c) of f fliers.
   Equation: _____

10) A hospital patient is allowed 2000 calories per day. The calories allowed for lunch are twice the calories allowed for breakfast. The calories allowed for dinner are $1\frac{1}{4}$ the calories allowed for lunch.

    a) Represent algebraically the number of calories in each meal.
    Breakfast: _____ Lunch: _____ Dinner: _____

    b) Set up an equation to solve for the number of calories allowed for each meal.
    Equation: _____

CLOSURE

Here are the solutions to the class fundraiser problem on page 103.

The cost of the fruit is $140.00 plus $0.55 for each pound sold over 125.

1) Algebraically, let C = the total cost of the fruit and f = the total number of pounds sold (assuming that f exceeds 125 pounds). We know the total cost of the fruit is made up of two parts:
   the cost of the first 125 pounds ($140.00) + the cost of the fruit in excess of 125 pounds
   To algebraically represent the cost of the fruit in excess of 125 pounds, we need two expressions:
   - the amount of fruit over 125 pounds represented by $f - 125$
   - the cost represented by $\$0.55 \times (f - 125)$ or $0.55(f - 125)$

   Now we are ready to set up the equation.

   Equation: $C = \$140.00 + \$0.55(f - 125)$

2) Use the formula we just developed to find the cost of 565 pounds of fruit.

   Equation: $C = \$140.00 + \$0.55(565 - 125)$

   Solution:
   $C = \$140.00 + \$0.55(440)$
   $C = \$140.00 + \$242.00$
   $C = \$382.00$

3) profit = total sales ($638.00) – total cost ($382.00). The class made **$256.00** profit from the fruit sale.

Good job!

Meeting the Challenge...Mathematics

LINEAR EQUATIONS, INEQUALITIES AND SYSTEMS

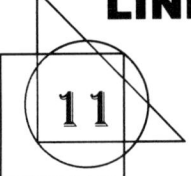

11 Use linear equations, inequalities, and systems to solve application problems.

FOCUS: Everyday activities present constant opportunities for problem solving. Sometimes we find ourselves developing our own formulas (equations) to find missing information.

Brandon, Celia and Kyle own a lawn care business. On one job they earned $90. Brandon worked twice as many hours as Celia, and Kyle worked three times as many hours as Celia. Their hourly pay rates are the same.
  1. How should the money be divided among Brandon, Celia and Kyle? Create an equation.
  2. If Celia worked 2.5 hours, what was her hourly rate for this job?
  3. Did Brandon make enough money to spend $7.95 to rent video games?
  4. Kyle needs to make at least $6.00 an hour when he works. Did he achieve this goal?

PURPOSE: When you have completed this lesson, you will be able to represent a problem situation as a linear equation or system with two variables, or an inequality, then solve.

WHAT YOU NEED TO KNOW

MATHEMATICS AND PROBLEM SOLVING VOCABULARY:

1. Distributive property
2. Equivalent expressions
3. Formula
4. Inequality
5. Linear equation
6. Linear system
7. Opposite
8. Ordered pair
9. Reciprocal
10. Similar or like terms

MEMORIZE THESE FACTS:

1. Properties of Equality create new equations for use in solving linear equations. Each new equation is equivalent to the original equation.
    - Addition Property of Equality: Add the same real number to both sides of an equation to create a new equivalent equation.
    - Subtraction Property of Equality: Subtract the same real number from both sides of an equation to create a new equivalent equation.
    - Multiplication Property of Equality: Multiply each side of an equation by the same real number to create a new equivalent equation.
    - Division Property of Equality: Divide each side of an equation by the same nonzero real number to create a new equivalent equation.
2. Use the opposite to subtract positive and negative numbers. To subtract a number, add its opposite.
3. Symbols used with inequalities:
    a. $<$, less than
    b. $\leq$, less than or equal to
    c. $>$, greater than
    d. $\geq$, greater than or equal to
    e. $\neq$, not equal to

© 2003 Orange Frazer Press. All Rights Reserved.

4. **Properties of Inequalities:** Except for one primary change, the Properties of Equality used to solve linear equations are used to solve inequalities. When multiplying or dividing both sides of the inequality by a negative value, the inequality sign reverses direction.
5. The solution of a linear system of equations falls into one of three categories.
    Case 1: There is a common ordered pair solution for the system.
    Case 2: Both variables are eliminated leaving a **true** statement. The solutions of this system are the ordered pairs that are solutions to either equation.
    Case 3: Both variables are eliminated leaving a **false** statement. There are no common solutions.

BE ABLE TO PERFORM THESE OPERATIONS:

Solving Linear Equations

1. **Use Properties of Equality to solve one-step equations.**
    - Remember that subtracting a number is the same as adding its opposite.
    - Remember that multiplying the coefficient of the variable by its reciprocal gives a result of 1.

    Examples:

    a) $f + 7 = 4$
    $f + 7 - 7 = 4 - 7$
    $f = -3$

    b) $a - (-3) = 9$
    $a + 3 = 9$
    $a + 3 - 3 = 9 - 3$
    $a = 6$

    c) $-6x = 24$
    $\dfrac{-6x}{-6} = \dfrac{24}{-6}$
    $x = -4$

    d) $\dfrac{x}{-3} = 7$
    $\dfrac{-3}{1} \cdot \dfrac{x}{-3} = 7 \cdot \dfrac{-3}{1}$
    $x = -21$

2. **Use more than one Property of Equality to solve equations.** Follow these steps:
    Step 1: Use the distributive property to remove all parentheses.
    Step 2: Combine like terms on each side of the equal sign.
    Step 3: Use addition or subtraction to move numerical terms to one side of the equation and variable terms to the other side of the equation.
    Step 4: Use multiplication or division to solve for the variable.

    Examples:

    a) $5x + 2 = -8$
    $5x + 2 - 2 = -8 - 2$
    $5x = -10$
    $\dfrac{5x}{5} = \dfrac{-10}{5}$
    $x = -2$

    b) $x - 4 = 5 - 2x$
    $x - 4 + 2x = 5 - 2x + 2x$
    $3x - 4 = 5$
    $3x - 4 + 4 = 5 + 4$
    $3x = 9$
    $\dfrac{3x}{3} = \dfrac{9}{3}$
    $x = 3$

    c) $3(2x - 4) - (3x - 4) = 1$
    $6x - 12 - 3x + 4 = 1$
    $3x - 8 = 1$
    $3x - 8 + 8 = 1 + 8$
    $3x = 9$
    $\dfrac{3x}{3} = \dfrac{9}{3}$
    $x = 3$

Solving Inequalities

Use Properties of Inequalities to solve inequalities. Remember: When multiplying or dividing both sides of an inequality by a negative value, the inequality sign reverses direction.

Examples:

a) $x + 5 \geq -9$
$x + 5 - 5 \geq -9 - 5$
$x \geq -14$

b) $-2x + 3 < 9$
$-2x + 3 - 3 < 9 - 3$
$-2x < 6$
$\dfrac{-2x}{-2} < \dfrac{6}{-2}$
$x > -3$

c) $\dfrac{x}{-4} > 12$
$\dfrac{x}{-4} \cdot \left(\dfrac{-4}{1}\right) > 12 \cdot \left(\dfrac{-4}{1}\right)$
$x < -48$

Meeting the Challenge...Mathematics

Solving Inequality Application Problems

1. **Write inequality expressions from word phrases.**
    - Word phrases: no greater than, no more than, at most, maximum (Inequality used: $\leq$)
      Example: A number is no greater than, no more than, at most, a maximum of 7. **Inequality:** $x \leq 7$
    - Word phrases: at least, no less than, minimum (Inequality symbol used: $\geq$)
      Example: A number is at least, no less than, or a minimum of 12. **Inequality:** $x \geq 12$

2. **To solve applications, follow these steps to write an equation or an inequality statement.**
    Step 1: Identify a word sentence that describes the mathematical relationship.
    Step 2: Convert the ideas given in words to algebraic terms.
    Step 3: Translate the word sentence to an algebraic equation or inequality.
    Example: In the first eleven months of the year, Jason earned $4375, but in the twelfth month he earned enough to make his year's income exceed $5100. What did he earn in the twelfth month?
    Representation: Let m = the money Jason earned the last month of the year.
    Then $m + 4375$ represents the earnings for the year. **Answer:** $m + 4375 > 5100$

3. **Follow these steps to solve a problem involving an algebraic formula.**
    Step 1: Replace values in the formula with known values. Be sure the units of given values match up.
    In $d = r \cdot t$, if r is given in miles per hour, the value of t must be in hours or changed to hours.
    Step 2: Use the Properties of Equality to solve the equation for the missing variable.
    Step 3: Verify that the answer is reasonable and that the units are correct.
    Example: The formula for the perimeter of a rectangle is $P = 2l + 2w$. If $P = 48$ and $l = 15$, find w.
    Substitute values: $48 = 2 \cdot 15 + 2w$ Solution: $48 = 30 + 2w$ $18 = 2w$ **Answer:** $w = 9$

Solving Linear Systems of Equations

Follow these steps to solve a linear system of two equations.
Step 1: Write both equations in standard form. ($ax + by = c$)
Step 2: Multiply one or both equations by numbers so the coefficients of one variable will be opposites.
Step 3: Add or subtract the equations from Step 2 to eliminate one of the variables.
Step 4: Solve the equation from Step 3 for the remaining variable.
Step 5: Substitute the value from Step 4 into one of the given equations. Solve for the other variable.
Step 6: To check the solution, substitute the values into one of the original equations.
Examples:

a) $2a + b = 12$ (Equation 1)
    $a - b = -3$ (Equation 2)
    $3a = 9$ (Add both)
    $a = 3$
Substitute $a = 3$ into one of the original equations. We'll choose #2.
    $3 - b = -3$
    $-b = -6$
    $b = 6$
    Solution: $(3, 6)$
To check, substitute $(3,6)$ into equation 1. $2(3) + 6 = 12$ ✓

b) $x - 4y = 14$ (Equation 1)
    $3x + y = 16$ (Equation 2)
    $x - 4y = 14$ (Bring down Eq. 1)
    $12x + 4y = 64$ ($4 \times$ Eq. 2)
    $13x = 78$ (Add both)
    $x = 6$
Substitute $x = 6$ into one of the original equations. We'll choose #2.
    $18 + y = 16$
    $y = -2$
    Solution: $(6, -2)$
To check, substitute $(6, -2)$ into equation 1. $6 - 4(-2) = 14$ ✓

c) $3x - 4y = 41$ (Equation 1)
    $7x + 6y = 19$ (Equation 2)
    $9x - 12y = 123$ ($3 \times$ Eq. 1)
    $14x + 12y = 38$ ($2 \times$ Eq. 2)
    $23x = 161$ (Add both)
    $x = 7$
Substitute $x = 7$ into one of the original equations. We'll choose #2.
    $49 + 6y = 19$
    $y = -5$
    Solution: $(7, -5)$
To check, substitute $(7, -5)$ into equation 1. $3(7) - 4(-5) = 41$ ✓

Meeting the Challenge...Mathematics

BE ABLE TO PERFORM THESE CALCULATOR OPERATIONS:

1. Use the positive/negative key $\boxed{+/-}$. Enter a number. Use this key to change the sign (take the opposite).
2. Use the calculator to verify solutions.

GUIDED PRACTICE

A) Solve these one-step equations.

1) $x + 2 = -7$
2) $x + 0.41 = 2.8$
3) $y - 8 = -12$
4) $y - \frac{2}{3} = 1\frac{2}{3}$

5) $-3x = -24$
6) $-x = 9$
7) $\frac{y}{-2} = 8$
8) $\frac{y}{2.6} = 1\frac{3}{4}$

B) Solve these two-step equations.

9) $2a - 7 = 9$
10) $4b - 3 = 13$
11) $5x + 6 = 21$
12) $3x + 5 = -8$

13) $\frac{4}{5}d - 4 = 20$
14) $\frac{3}{4}x + 13 = 7$
15) $7x + 1.3 = 36.3$
16) $3d + 1\frac{1}{3} = -6\frac{2}{3}$

C) Solve these multiple-step equations.

17) $2x + 20 = 80 - 4x$
18) $7x + 13 = 5x - 20$
19) $2(3g - 5) = 4g + 12$

20) $2k - 9 = 7(2k - 3)$
21) $5b - 2(2b + 3) = 4 - b$
22) $4m + 3m - 2m + 1 = 3 - m + 4$

Meeting the Challenge...Mathematics

D) Solve these one-step inequalities.

23) $x + 6 \geq -3$ 24) $t - 5 > -11$ 25) $-5d \leq 35$ 26) $8p > -64$ 27) $-\dfrac{y}{8} \geq -3$

E) Solve these multiple-step inequalities.

28) $5c - 6 \geq 9$ 29) $-3y + 2 > -10$ 30) $y + 1 < 7 + 3y$ 31) $4 - 2x \leq -5 - x$

F) Fill in the correct inequality symbol to complete the expression.

32) A number is no greater than 55. $n \;\boxed{}\; 55$ 33) A number is at least 16. $n \;\boxed{}\; 16$

G) Find the missing value in the formulas by substituting the given values.

34) $V = l \bullet w \bullet h$ $V = 252, \; l = 7, \; w = 12$ 35) $D = r \bullet t$ $D = 220, \; r = 55$

$h =$ _____ $t =$ _____

H) Write the algebraic equation that expresses the statement.

36) The number of boys in a school is twice the number of girls. The total enrollment of the school is 615 students. Represent algebraically.

a) the number of boys: _____

b) the number of girls: _____

c) the equation to solve for the number of boys and girls enrolled:
_____

37) The junior class sponsored a donkey basketball game to raise money for the local homeless. Five hundred tickets were sold. Adult tickets were $4 and student tickets were $2. The school raised $1600 for the shelter. Represent algebraically.

a) the number of adult tickets sold: _____

b) the number of student tickets sold: _____

c) the total number of student and adult tickets sold: _____

d) the equation to solve for the number of adult and student tickets sold:
_____

© 2003 Orange Frazer Press. All Rights Reserved.

I) Find the solutions for the linear systems.

38) $\begin{aligned} x - y &= 1 \\ x + y &= 9 \end{aligned}$

39) $\begin{aligned} 4x - 10y &= 0 \\ 2x + y &= 12 \end{aligned}$

40) $\begin{aligned} 6x - 5y &= 33 \\ 4x + 4y &= 44 \end{aligned}$

x = _____ y = _____

x = _____ y = _____

x = _____ y = _____

INDEPENDENT PRACTICE

A) Solve and simplify answers.

1) $n + 81 = 97$

2) $\frac{3}{4}a = 27$

3) $y + \frac{5}{8} = \frac{13}{16}$

4) $c - \frac{7}{8} = 1\frac{5}{8}$

5) $\frac{d}{5} = 12$

6) $m + 5 = -6.9$

7) $50 = a - 10$

8) $x - 0.19 = 0.3$

9) $45 = \frac{5}{6}n$

10) $\frac{y}{8} = -7$

11) $a + 2.45 = 6$

12) $f - 4 = -2.5$

13) $5g + 2g + g = 104$

14) $9x - 7 - 5x = 13$

15) $6.4x - x + 3.6x = 135$

16) $-a - 10 + 6a = 40$

17) $-12 - b + 4b = -18$

18) $x - 4 = 5 - 2x$

19) $8a + 1 = 10 + 4a$

20) $8 - y = 14 - 2y$

21) $2.9x - 6.6 = 27.8 - 1.4x$

Meeting the Challenge...Mathematics

22) $3(2x - 4) - (3x - 4) = 1$

23) $5y - 2(2y + 3) = 4 - y$

24) $5(3x - 2) - 3(4x - 1) = 5$

25) $7(2 - y) - (8 - 5y) + y = 0$

26) $5y - 18 = 5(2y - 3) - 4y$

27) $3b - 16 = 8b - 6(b + 1)$

28) $18 - (4x + 9) = 3(2 - x) + 4$

29) $12a - (6a + 20) = 8a - 4(2a - 5)$

B) Express each application word problem as an equation and solve.

30) A length of pipe 40 feet long is cut into two different lengths. One piece is four times the length of the other. Find the lengths of the two pieces of pipe.

a) Give the algebraic representation for both lengths of pipe.
   _____ , _____

b) Write an equation and solve.
   Equation: _____

   Solution:

c) The lengths of the two pieces are
   _____ and _____ .

31) There are 21 more girls than boys in a graduating class of 271 students. How many boys are in the class? Girls?

a) Give the algebraic representation for each.
   boys _____

   girls _____

b) Write an equation and solve.
   Equation: _____

   Solution:

c) There are _____ boys and _____ girls in the class.

32) In a banquet hall, it costs five times as much to heat the entire seating area as it does to heat the serving area. If it costs $138 to heat the banquet hall for an event, find the cost to heat the seating area and the cost to heat the serving area.

a) Give the algebraic representation for the cost to heat the areas.

seating area _____

serving area _____

b) Give an equation and solve.

Equation: _____
Solution:

c) It costs _____ to heat the seating area

and _____ to heat the serving area.

33) During the year (not leap year), Carmen's hometown had 6 fewer days of clear weather than of cloudy weather and 4 more days of clear weather than of partly cloudy weather. Find the number of days of each kind of weather.

a) Give an algebraic representation for the types of weather.
clear _____ cloudy _____

partly cloudy _____

b) Give an equation and solve.

Equation: _____
Solution:

c) During the year there were _____ days of clear weather, _____ days of cloudy weather, and _____ days of partly cloudy weather.

C) Find the missing value in each formula.

34) Formula: $d = mw$

$d = 72$ and $m = 9$

$w =$ _____

35) Formula: $F = 1.8C + 32$

$F = 179.6$

$C =$ _____

36) Formula: $I = \dfrac{E}{R}$

$I = \dfrac{3}{4}$ and $R = 92$

$E =$ _____

D) Solve the inequalities.

37) $y - 5 > -12$

38) $-6x \leq -54$

39) $a + \dfrac{3}{5} \geq -4\dfrac{4}{5}$

40) $-1.5x \leq -6$

41) $b + 7 > -10$

42) $-10 < -2.5d$

43) $5x - 1 \leq 19$

44) $6 - 3x \geq 24$

45) $6x + 2 \leq 26 - 2x$

46) $7y - 4 > -12 + y$

47) $-3(2y - 8) > 2(3 + y)$

Meeting the Challenge...Mathematics

E) **Write an inequality for each word sentence.**

48) A number is greater than –4.   Inequality: _____

49) Five times a number is no less than 9.   Inequality: _____

50) A number decreased by 11 is less than 36.   Inequality: _____

51) A number is at least 7 greater than twice the number.   Inequality: _____

F) **Express each application word problem as an inequality and solve.**

52) The budget allows for no more than $125.00 to be spent on printing fliers for the school fair. The fliers will cost $65.00 for the first 1000 and $0.05 a copy for each flier over 1000.

   a) Set up an inequality to find the maximum number of fliers the school can buy and still stay within the budget amount.   Inequality: _____

   b) Solve the inequality.   Maximum number of fliers: _____

53) Antonio is responsible for paying for repairs to the family car after a minor accident. The materials will cost at least $1525 and labor will cost $48 per hour.

   a) Give an algebraic inequality to calculate the cost of getting the car fixed. Use c to represent the cost of getting the car fixed in dollars and h to represent the number of hours of labor.

   Inequality: _____

   b) The repairman estimates that it will take 7 hours to complete the repairs. Solve the inequality for c.   Answer: _____

G) **Solve the linear systems.**

54) $2c + d = 12$
    $c - d = -3$

55) $3a - b = 11$
    $a - b = 5$

56) $x - 4y = 14$
    $3x + y = 16$

57) $4x + 5y = 7$
    $3x + 4y = 4$

$c =$ ___ $d =$ ___ $a =$ ___ $b =$ ___ $x =$ ___ $y =$ ___ $x =$ ___ $y =$ ___

H) Use a linear system of equations to solve the application problems.

58) Three hundred thirty-six people attended the school's fall spaghetti dinner. Student tickets were $4 and adult tickets were $7. The school collected a total of $1959.

   a) Set up a linear system of equations to find the number of student tickets (s) and adult tickets (a) sold.

   System: _____   Number of adult tickets sold = _____

   _____   Number of student tickets sold = _____

   b) Solve.

59) Mrs. Guthrie owns ten one and two bedroom apartments. The one bedroom apartments rent for $300 per month and the two bedroom apartments rent for $400 per month. In March, she collected a total of $3700 for the ten apartments.

   a) Set up a linear system of equations to find the number of one bedroom apartments (x) and two bedroom apartments (y).

   System: _____   Number of one bedroom apartments = _____

   _____   Number of two bedroom apartments = _____

   b) Solve.

TEST PRACTICE

MULTIPLE–CHOICE QUESTIONS: Write the letter of the correct answer.

1) Solve. $2x + 20 = 80 - 4x$ _____

   a. -50 b. 10 c. -30 d. 50

2) If $S = \frac{n}{2}(a + l)$, what is the value of a when $S = 48$, $n = 4$ and $l = 5$?   _____

   a. 21.5 b. 38 c. 19 d. 46

Meeting the Challenge...Mathematics

3) Three times a number decreased by 8 gives the same result as twice the number added to 12. Find the number.
    a. 20   b. 4   c. −4   d. −20

4) Find the solution for x in the following inequality: $-5x - 1 \leq x + 17$
    a. $x \leq -2\frac{2}{3}$ b. $x \leq -3$ c. $x \geq -3$ d. $x \geq -2\frac{2}{3}$

5) An item is on sale for 25% off the original price. Which equation can be used to find the original price if the sale price is $36?
    a. $x + 0.25x = 36$ b. $x + 36x = 0.25$ c. $0.25x = 36$ d. $x - 0.25x = 36$

SHORT ANSWER QUESTIONS:

6) Julian ordered a total of 25 hamburgers and cheeseburgers. The price of a hamburger is $2.75 and the price of a cheeseburger is $2.95.

   a) To solve this problem using only one equation, the algebraic representation for the number of hamburgers is _____ and for the number of cheeseburgers is _____ .

   b) To solve this problem using only one equation, the algebraic representation for the cost of the hamburgers is _____ and for the cost of the cheeseburgers is _____ .

7) Refer to the information in problem 6. Julian spent $70.55 for the sandwiches. Write the equation and solve for the number of hamburgers and cheeseburgers that Julian purchased.

   Equation: _____   Solve.
   Hamburgers: _____
   Cheeseburgers: _____

EXTENDED RESPONSE QUESTIONS:

8) Solve the linear system. $5x + 6y = 9$
                            $3x - 4y = 13$

   Answers:

   $x = $ _____

   $y = $ _____

9) As a fundraiser during the holidays, the Drama Club wrapped packages at the mall. They charged $2.50 for a small package and $4.00 for a large package. They wrapped a total of 658 packages and collected $2269.00 for the club.

   a) Set up an equation to solve for the number of small and large packages.

   Equation: _____

   b) Solve.

   Answers: The number of small packages was _____, and the number of large packages was _____.

10) Is $x = 6$ and $y = 2$ the solution to the linear system of equations, $x - y = 14$ and $x + y = 8$?

   Yes ___   No ___

   Justify your answer.

CLOSURE

Let's return to the Focus problem on page 117. We will use algebra to help the three friends answer their business related questions.

1) Since each person makes the same hourly rate, we can represent algebraically the money each earned based on the number of hours worked. Let x = the amount of money Celia earned. Therefore, $2x$ = the amount Brandon earned and $3x$ = the amount that Kyle earned.
   - The sum of these three amounts equals the total amount they earned which was $90.00. Algebraically, the equation to use is $2x + x + 3x = \$90$ $6x = \$90$ $x = \$15$
   - This value of x represents what Celia earned. Brandon earned double this amount or $30, and Kyle earned three times this amount or $45.

2) Celia earned $15 for 2.5 hours of work. Divide the amount earned by the number of hours worked to solve for her hourly rate. Calculate: $15 \div 2.5$ Answer: $6 per hour

3) Brandon earned $30 which gives him more than enough to rent the $7.95 video game.

4) Did Kyle end up making $6.00 an hour? All three workers made the same hourly wage. So, Kyle earned exactly $6.00 per hour, the same as Celia. Brandon, Celia and Kyle earned some money and had fun, too!

Meeting the Challenge...Mathematics

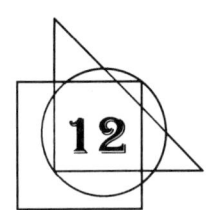

CREATING AND ANALYZING GRAPHS

Create, analyze and apply linear and nonlinear graphs.

FOCUS: Let's explore using algebraic graphs in real world situations.

Your high school choir wants to purchase personalized t-shirts to wear for special events. The cost of the t-shirt with the school's name printed on the back is $8. Each member's first name will be printed on the front for an additional $0.50 per letter.
1. Develop a formula to find the total price of a t-shirt.
2. Use your formula to find the t-shirt price for Charlene.
3. Draw a graph to represent the price of a t-shirt with up to twelve letters in the first name. Represent the price of the t-shirt on the vertical axis or *y*-axis and the number of letters in the first name on the horizontal or *x*-axis.
4. Use your graph to find the t-shirt price for Charlene.
5. Did you get the same price for Charlene's t-shirt using the formula and the graph?
6. With your formula in slope-intercept form, tell the slope and explain what it means in terms of this problem.
7. From your graph, give the *y*-intercept and explain what it means in terms of this problem.

PURPOSE: When you have completed this lesson, you will be able to graph and analyze linear equations using several techniques and apply your observations to real world situations. You will also be able to identify linear and nonlinear relationships.

WHAT YOU NEED TO KNOW

MATHEMATICS AND PROBLEM SOLVING VOCABULARY:

1. Constant
2. Coordinates of a point
3. Degree of an equation
4. Domain
5. Horizontal or *x*-axis
6. Linear equation
7. Nonlinear function
8. Ordered pair
9. Origin
10. Quadrants
11. Quadratic equation
12. Range
13. Roots of an equation or function
14. Scatter plot
15. Slope of a line
16. Vertical or *y*-axis
17. Y-intercept of an equation
18. Zeros of an equation

MEMORIZE THESE FACTS AND FORMULAS:

1. The horizontal axis and vertical axis divide the graph into four parts or quadrants. The properties of these quadrants are listed below.
   - 1st quadrant: *x* and *y* are both positive.
   - 2nd quadrant: *x* is negative and *y* is positive.
   - 3rd quadrant: *x* and *y* are both negative.
   - 4th quadrant: *x* is positive and *y* is negative.

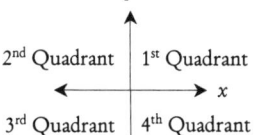

© 2003 Orange Frazer Press. All Rights Reserved.

2) The slope of a line = $\frac{\text{vertical change}}{\text{horizontal change}}$ or $\frac{\text{rise}}{\text{run}}$. $F = (x_1, y_1)$ $G = (x_2, y_2)$ slope of $\overline{FG} = \frac{y_2 - y_1}{x_2 - x_1}$

3) The slope-intercept form of the equation of a line is: $y = mx + b$ (m = slope, b = y-intercept)

4) There are four possible line/slope relationships that come from the value of the slope.
   Relationship 1: A line with a positive slope rises from left to right. Example: walking up a hill
   Relationship 2: A line with a negative slope falls from left to right. Example: walking down a hill
   Relationship 3: A line with a zero slope (rise = 0) is horizontal. Example: walking on level ground
   Relationship 4: A line with no defined slope (run = 0) is vertical. Example: scaling a wall

5) The equation of a horizontal line is of the form $y = k$ (a constant number).

6) The equation of a vertical line is of the form $x = h$ (a constant number).

7) The zeros, roots or solutions of an equation or function are found by setting the expression equal to zero and solving for all values of x.

8) Any equation or function that cannot be converted into slope-intercept form ($y = mx + b$) is not linear.

BE ABLE TO PERFORM THESE OPERATIONS:

1. Plot points on the x-y coordinate graph.
   Examples: Answer:
   Plot these points.

   $A = (4, -3)$ $C = (2, 1)$ $E = (0, -3)$

   $B = (-3, -3)$ $D = (-2, 4)$ $F = (3, 0)$

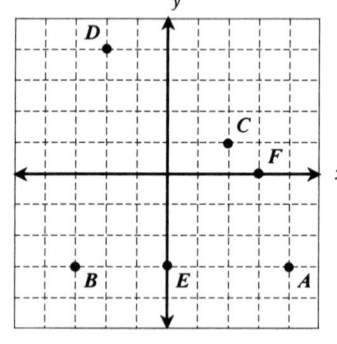

2. Convert the equation of a line to slope-intercept form by solving for y.
   Example: $2x + 3y = 9$
   Solution: $2x + 3y + (-2x) = 9 + (-2x)$ $3y = -2x + 9$ Answer: $y = -\frac{2}{3}x + 3$

3. Find the slope of a line.
   Case 1: The coordinates of two points on the line are known. Use the formula for slope. $m = \frac{y_2 - y_1}{x_2 - x_1}$

   Example: $J = (3, -2)$ $K = (5, -4)$

   $m = \frac{-4 - (-2)}{5 - 3} = \frac{-4 + 2}{2} = \frac{-2}{2} = -1$
   Answer: $m = -1$

   Case 2: The slope-intercept form of the line is given. Use the slope-intercept form of the equation. $y = mx + b$

   Example: $y = \frac{4}{3}x + 2$

   Answer: $m = \frac{4}{3}$

Meeting the Challenge...Mathematics

Case 3: The equation is not in slope-intercept form. Solve for y and put the equation in slope-intercept form.
Example:
$4x + 5y = 20$
$5y = -4x + 20$
$y = -\frac{4}{5}x + 4$
Answer: $m = -\frac{4}{5}$

Case 4: A linear graph is given.
Example:

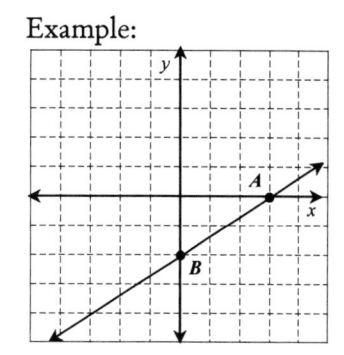

Solution: Identify and use the coordinates of the x and y intercepts to solve for the slope.
x-intercept or point A = (3, 0)
y-intercept or point B = (0, –2)
Apply the slope formula or use $\frac{rise}{run}$.
To go from point B to point A, the rise = 2 and the run = 3.
Answer: $m = \frac{2}{3}$

---

4. Graph a linear equation.

Case 1: Two points on the graph are given. Plot the points and draw the line.
Example: A = (3, 4), B = (–1, 1)

Solution: Plot the points and draw the line through them.

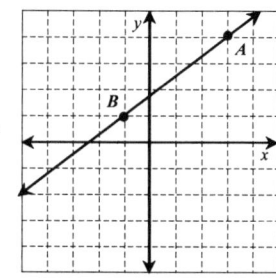

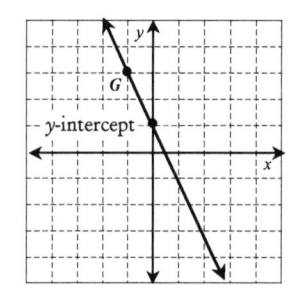

Case 2: The y-intercept and the slope of the line are given.
Example: y-intercept = 1 and $m = -2$
Solution: Plot the y-intercept at (0, 1).
Use $m = -\frac{2}{1} = \frac{rise}{run}$ to go up 2 and 1 to the left to point G (–1, 3).
Draw the line through the y-intercept (0,1) and (–1, 3).

Case 3: Graph a line when the equation of the line in slope-intercept form is given.

Example: $y = -\frac{3}{2}x - 1$

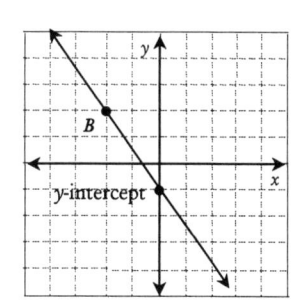

Solution: From the equation, the y-intercept is –1 and the slope is $-\frac{3}{2}$.

From the y-intercept of –1, go up 3 units and 2 units to the left to plot point B (–2, 2) on the graph. Draw the line through the y-intercept of –1 and point B.

Case 4: The coordinates of one point and the slope of the line are given.
Example: Graph the line containing point C(–3, –2) with slope = $\frac{3}{5}$.

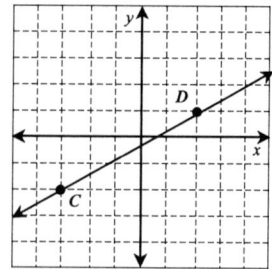

Solution: Plot point C. Find the coordinates of another point, D, on the line by using the slope. Find the y–coordinate of D by going up 3 units or adding 3 onto –2. The y–coordinate of D = 1. Find the x–coordinate of D by going to the right 5 units or adding 5 onto –3. The x–coordinate of D = 2. Plot point D(2, 1) and draw the line through point C.

Case 5: The equation of the line is not given in slope-intercept form. Convert the equation to slope-intercept form and then graph. 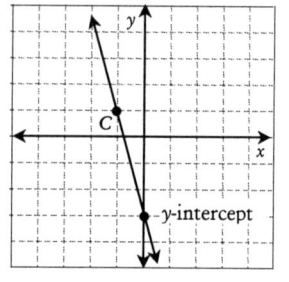
Example: $4x - y = 3$
Solution: $4x + (-4x) - y = 3 + (-4x)$ $-y = -4x + 3$
Now change each sign. Slope-intercept form: $y = 4x - 3$
From the y-intercept of –3, go up 4 units and go one unit to the left and plot point C (–1, 1).
Draw the line through the y-intercept of –3 and (–1, 1) to graph the line.

Case 6: The equation of the line in the form $x = h$ (a constant number) is a vertical line on that x-coordinate. The equation of the line in the form $y = k$ (a constant number) is a horizontal line on that y-coordinate.

a) Example of a vertical line: $x = 3$
  Every point on this line has 3 as its x-coordinate.

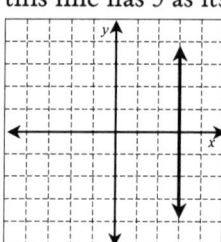

b) Example of a horizontal line: $y = -2$
  Every point on this line has –2 as its y-coordinate.

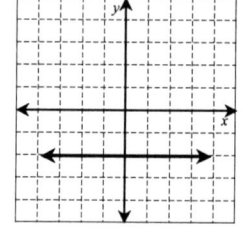

5. **Determine the equation of a line in slope-intercept form.** From the information given, solve for the y-intercept and the slope. Substitute m and b into the slope-intercept form.

   Case 1: A point on the line and the slope is given. Solve for the y-intercept and then substitute into the equation.
   Example: Point A = (3, –1), slope = $-\frac{1}{5}$
   Solution: To solve for b, substitute (3, –1) and $-\frac{1}{5}$ into the slope-intercept form, $y = mx + b$.
   Solve for b: $-1 = \left(-\frac{1}{5}\right)(3) + b$ $-1 = \left(-\frac{3}{5}\right) + b$ $-\frac{2}{5} = b$ Answer: $y = -\frac{1}{5}x - \frac{2}{5}$

   Case 2: The coordinates of two points are given. Solve for the slope. Substitute the slope and the coordinates from one of the given points into $y = mx + b$ to solve for b.
   Example: R = (3, –2) S = (–1, 4)
   Solution: Slope = $\frac{4 - (-2)}{-1 - 3} = \frac{6}{-4} = -\frac{3}{2}$
   Use the coordinates from either R or S to substitute into $y = mx + b$ in order to solve for b.
   Using R, $-2 = -\frac{3}{2}(3) + b$ $-2 = -\frac{9}{2} + b$ $\frac{5}{2} = b$ Answer: $y = -\frac{3}{2}x + \frac{5}{2}$

Meeting the Challenge...Mathematics

Case 3: The graph is given. Look at the graph to find the x and y intercepts. Then use these two points to find the slope.

Example: Solution: From the graph, x-intercept = $(-3, 0)$ and the y-intercept = $(0, -4)$

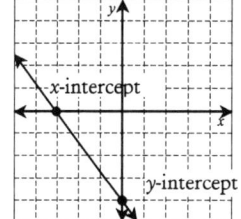

$$\text{Slope} = \frac{-4-0}{0-(-3)} = -\frac{4}{3}$$

Answer: $y = -\frac{4}{3}x - 4$

6. **Determine whether a graph is linear or nonlinear.** The equation of a line can be expressed in slope-intercept form, $y = mx + b$, where both x and y have exponents of one. This equation is said to have degree one. Any equation or function that cannot be converted to slope-intercept form is nonlinear.
   Examples of nonlinear equations:
   a) $y = x^2 + 3x + 2$ has an exponent of 2.
   b) $y = x^3 + 2$ has an exponent of 3.
   c) $y = \frac{4}{x}$ cannot be written in slope-intercept form.

7. **Find the roots, zeros or solution of a function or equation from its graph.** These values are the points where the graph crosses the x-axis. Look at the graph in example #8.
   Answer: The graph crosses the x-axis at $(-1, 0)$ and $(3, 0)$. These two points are the zeros, roots or solutions of the equation, $y = x^2 - 2x - 3$.

8. **Plot points to determine if an equation is linear or nonlinear.** Use this method to verify whether an equation is linear or nonlinear if you can't decide by looking at the equation itself.
   Step 1: Develop a table of coordinates. Choose basic positive and negative values and zero for x (such as $x = -2, -1, 0, 1, 2$). Substitute these values into the equation to find the corresponding values for y.
   Step 2: Graph the coordinate pairs. Note: When $x = 0$, the point is a y-intercept. When $y = 0$, x is a root or solution.
   Step 3: Choose more values, if necessary, to get a good overall view of the graph.
   Step 4: Connect the points and decide whether the graph is linear or nonlinear.

Example: Graph: $y = x^2 - 2x - 3$
Step 1: Make a table.   *Step 2*: Graph the coordinate pairs.

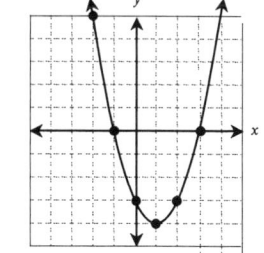

x	−2	−1	0	1	2	3
y	5	0	−3	−4	−3	0

Step 3: Add $x = 3$ to the table, and graph its point $(3, 0)$ to get a clearer idea of the shape of the graph.

Step 4: The graph is a nonlinear equation.

Meeting the Challenge...Mathematics

9. **Plot data points to determine if the points appear to have a linear relationship.** If so, draw a 'best fit' or trend line to represent the data, and to determine the slope, the y-intercept and the equation.
   Step 1: Plot the data on an x-y coordinate graph.
   Step 2: Sketch a 'best fit' line that closely follows the pattern of the points on the graph.
   Step 3: Pick two points on the line and determine their coordinates.
   Step 4: Find the slope.
   Step 5: Find the y-intercept.
   Step 6: Give the slope-intercept form of the equation.

   Example: Look at the points on the graph.
   Step 1:
     Plot data.
   Step 2: Sketch the 'best fit' line.
   Step 3: Label point A $(-1, 0)$ and point B $(1, 4)$.
   Step 4: From point A to B, the rise is 4, the run is 2. The slope is 2.
   Step 5: Use the line to approximate the y-intercept at $(0, 2)$.
   Step 6: Equation: $y = 2x + 2$

10. **Apply graphing concepts to real world situations.** Look for representations of slope and y-intercept in real-life application problems.
    Example: During the first year of satellite service in a small southeastern Ohio town, twelve units were installed. During the next three years, the number of satellite subscribers increased from 12 to 75. Look at the graphical representation of growth from the end of the 1st year to the end of the 4th year.

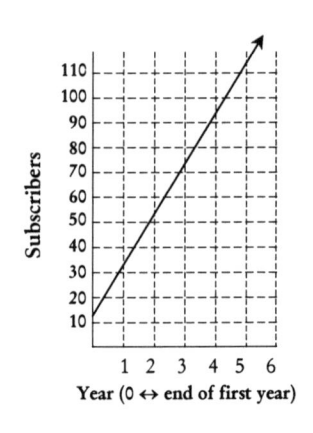

    a) Find the slope.   Answer: slope = $\frac{75-12}{3} = \frac{63}{3} = 21$

    b) Tell what the slope represents.   Answer: the yearly growth rate in satellite subscriptions

    c) Find the y-intercept.   Answer: 12

    d) Tell what the y-intercept represents.   Answer: the first year's sales

    e) If the sales continue to grow at the same rate, estimate the number of subscribers at the end of the 6th year.
       Answer: 100+ subscribers

BE ABLE TO PERFORM THESE CALCULATOR OPERATIONS:

1. Use the positive/negative key $\boxed{+/-}$. Enter a number. Use this key to change the sign (take the opposite).
2. Use the division key $\boxed{\div}$ to find the constant of variation of a ratio.
3. Use the square root key $\boxed{\sqrt{x}}$. Enter a number and use this key to find its square root.
4. Use the square key $\boxed{x^2}$. Enter a number and use this key to square it.

Meeting the Challenge...Mathematics

GUIDED PRACTICE

A) Use this graph for problems 1–5.

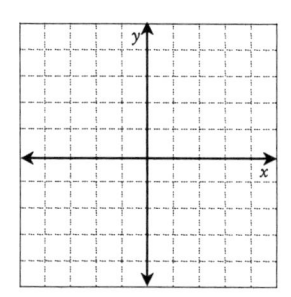

1) Plot these points and label them with the appropriate letter.
   A = (–5, 3)   C = (4, 3)   E = (2, 0)
   B = (3, –2)   D = (–3, –1) F = (0, –3)

2) Locate each point by the quadrant or another description.

Point	A	B	C	D	E	F
Location						

3) Moving from point A to point B, the rise is _____, the run is _____, making the slope _____.
4) Moving from point C to point D, the rise is _____, the run is _____, making the slope _____.
5) Moving from point E to point F, the rise is _____, the run is _____, making the slope _____.

B) Tell whether the following graphs have positive slope, negative slope, zero slope or no slope.

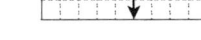

6) 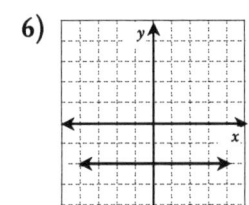 _____ slope   7) 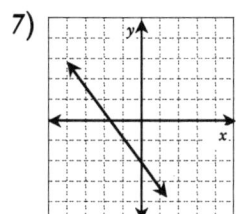 _____ slope   8) 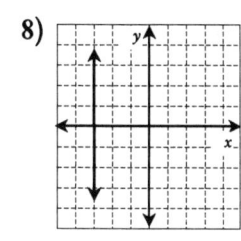 _____ slope   9) 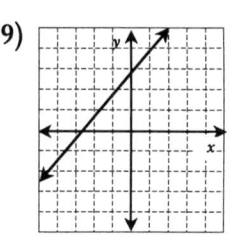 _____ slope

C) Match each letter with a graph.   a) $y = 3x + 2$ b) $y = -2x - 1$ c) $y = 3$ d) $x = 3$

10) _____   11) 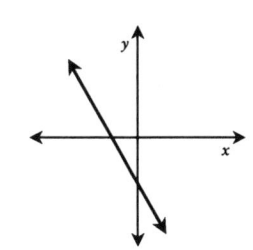 _____   12) 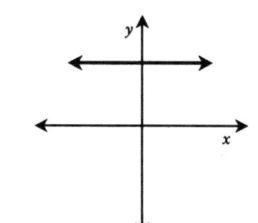 _____   13) _____

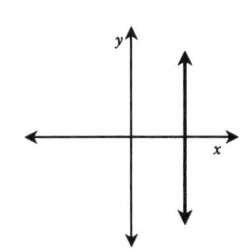

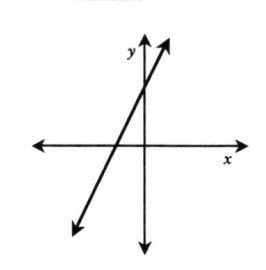

D) Use the equation to determine the slope.

Equation:	$y = x + 5$	$y = -3x + 1$	$y = -1$	$x = 5$
Slope:	14)	15)	16)	17)

© 2003 Orange Frazer Press. All Rights Reserved.

E) Convert each equation to slope-intercept form and give the slope.

18) $4x + y = 6$ 19) $x - 5y = 10$ 20) $3x - 4y = 20$ 21) $2x - y = 4$

Equation: _____ _____ _____ _____

Slope: ____ ____ ____ ____

F) Find the slope of the line that connects the following pairs of points.

22) M = (4, 1) N = (1, 6) slope = ____

23) A = (−2, −2) B = (5, −4) slope = ____

G) Use the rise and the run to determine the slope of the line on each graph.

24) x-intercept = ____
    y-intercept = ____
    rise = ____
    run = ____
    slope = ____

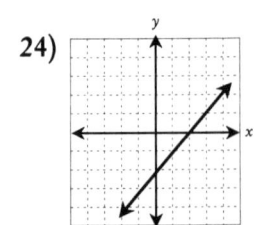

25) Label two points, R and S, on the line.
    rise = ____
    run = ____
    slope = ____

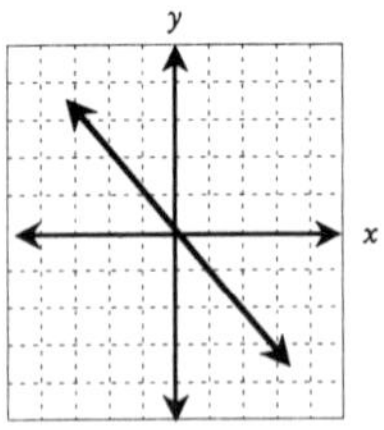

H) Graph the lines using the given information.

26) $m = \frac{2}{3}$, $b = -2$

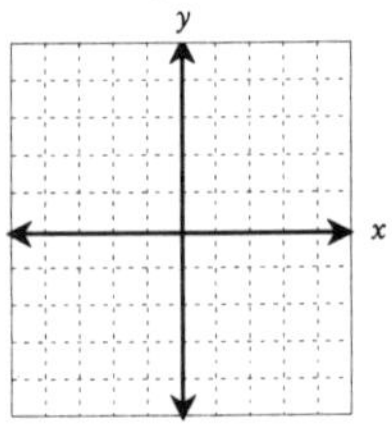

27) $m = -3$, containing (2, −2)

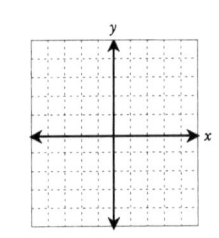

28) $y = -\frac{1}{2}x + 3$

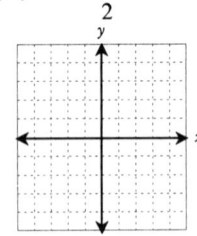

29) $2x - 5y = -10$

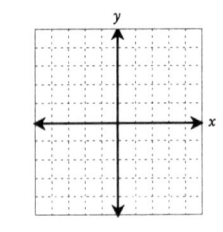

I) Find the slope-intercept form of the equation.

30) $m = \frac{3}{4}$ $b = -3$

31) $m = -3$, containing (9, 5) $b =$ ____

Equation: _____

Equation: _____

32) P = (5, 2) Q = (−1, 4)

33) Use the information from the graph.

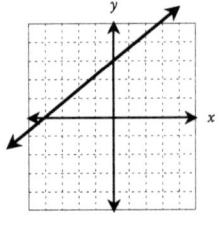

m = ____ b = ____

m = ____ b = ____

Equation: _____ Equation: _____

Meeting the Challenge...Mathematics

J) **Tell if the following equations are linear or nonlinear and explain your reasoning.**

34) $y = x^2 + 2x - 3$

linear or nonlinear (circle one)

Why? _____

35) $y = \dfrac{5}{x}$

linear or nonlinear (circle one)

Why? _____

36) $y = 4^x$

linear or nonlinear (circle one)

Why? _____

K) **Use the data on the scatter plot graph for the following questions.**

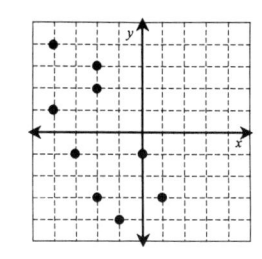

37) Draw a 'best fit' line.

38) The 'best fit' line crosses the x-axis at (,) and the y-axis at (,).

39) Determine the slope of the 'best fit' line. _____

40) Determine the equation of the line. _____

L) **Solve these real world applications of linear graphs.**

41) The high school jazz band must rent a van to travel to a concert performance. The rental company charges a flat fee of $55 plus 40¢ per mile.

  a) Write a linear equation that will find the total rental cost (y) in terms of the number of miles (x) driven. Equation: _____

  b) Complete the table.

Miles (x)	0	100	200	300	400
Cost (y)					

  c) Graph.

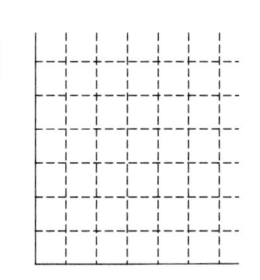

  d) What is the y-intercept and what does it represent in this problem?
  _____

  e) What is the slope and what does it represent in this problem?
  _____

  f) Use the graph to estimate the rental cost for a 350 mile trip.

  Cost = $ _____

42) Isaac works after school at the local hardware store. He receives a base pay of $400 each month plus a 5% commission on his sales.

  a) Write a linear equation that will calculate Isaac's total monthly pay (y) in terms of sales (x).
  Equation: _____

  b) Complete the table.

Sales $ (x)	0	200	400	600	800
Monthly pay (y)					

  c) Graph.

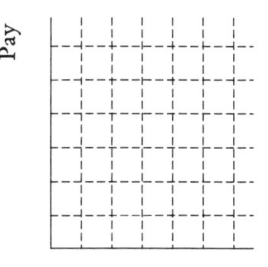

  d) What is the y-intercept and what does it represent in this problem?
  _____

  e) What is the slope and what does it represent in this problem?
  _____

  f) Use the graph to estimate Isaac's monthly pay based on sales of $500.

  Monthly pay = $ _____

© 2003 Orange Frazer Press. All Rights Reserved.

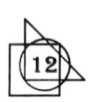

INDEPENDENT PRACTICE

A) Use this graph for the following problems.

1) Give the coordinates of the points and the location by quadrant or another description.

   M = (,) located _____   R = (,) located _____
   N = (,) located _____   S = (,) located _____
   P = (,) located _____   T = (,) located _____

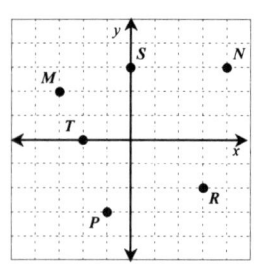

2) Tell if the slope of the line from one point to another is positive, negative, zero, or no slope.
   a) P to M _____   b) N to T _____   c) R to S _____
   d) T to the origin _____   e) the origin to M _____   f) N to the closest point on the x-axis _____

3) Moving from point R to point P, the rise is _____ and the run is _____. The slope = _____.
4) Moving from point T to point N, the rise is _____ and the run is _____. The slope = _____.
5) Moving from the origin to point M, the rise is _____ and the run is _____. The slope = _____.
6) Moving from point S to the origin, the rise is _____ and the run is _____. The slope = _____.

B) Tell whether the following lines have positive (a), negative (b), zero (c) or no slope (d).

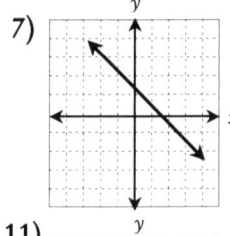

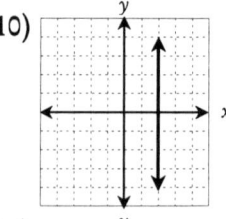

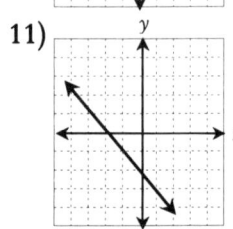

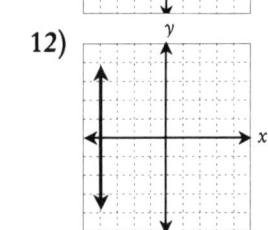

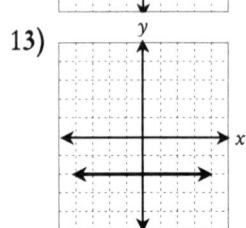

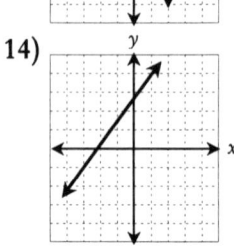

7)   8)   9)   10)

11)   12)   13)   14)

15) walking up the steps   16) driving down the ramp of the parking garage   17) boating on the lake

18) using an elevator   19) a decrease in sales over the last five years   20) 4 years of profit

Answers:
7) ___   8) ___   9) ___   10) ___   11) ___   12) ___   13) ___
14) ___ 15) ___ 16) ___ 17) ___   18) ___   19) ___   20) ___

138

Meeting the Challenge...Mathematics

C) Choose a graph that could represent the following equations.

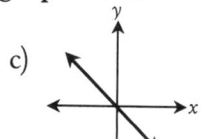

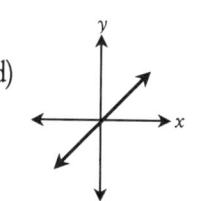

21) $y = 2x - 3$
22) $y = -2$
23) $y = -2x - 3$
24) $x = -2$
25) $y = \frac{3}{5}x + 1$
26) $y = -\frac{3}{5}x + 1$
27) $y = x$
28) $y = -x$

Answers:
21) ___
22) ___
23) ___
24) ___
25) ___
26) ___
27) ___
28) ___

D) Find the slope of the line from the given information.

29) $y = -4x + 3$
30) $y = x - 2$
31) containing $(-1, 3)$ and $(4, 7)$

32) $2x + y = 5$
33) $3x - 2y = 8$
34) containing $(2, -4)$ and $(-1, 3)$

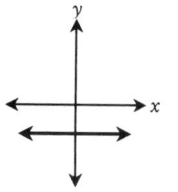

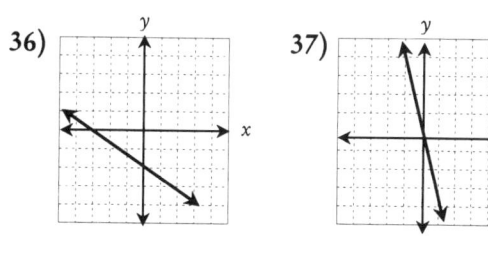

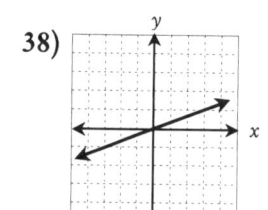

Answers:
29) _____
30) _____
31) _____
32) _____
33) _____
34) _____
35) _____
36) _____
37) _____
38) _____

35)   36)   37)   38)

E) Graph the lines using the given information. Give the equation of the line in slope-intercept form.

39) containing $(4, -1)$ and $(-3, 4)$
40) $m = 4, b = -3$
41) $5x + 3y = -9$
42) $m = 3$, containing $(-3, 2)$

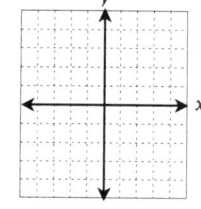

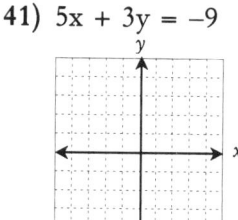

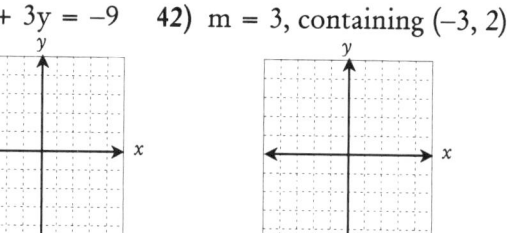

Equation:
_____   _____   _____   _____

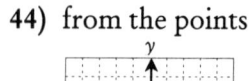

43) $x - 2y = 6$ 44) from the points 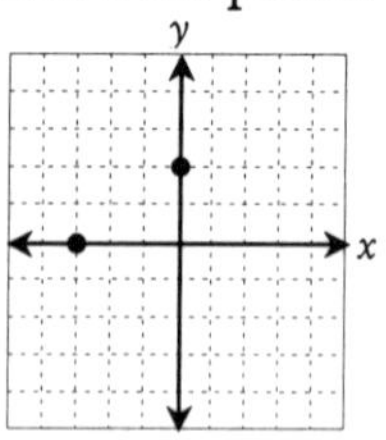   45) $m = -2, \ b = 1$ 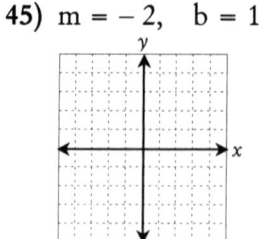   46) $m =$ no slope, containing $(-3, 2)$

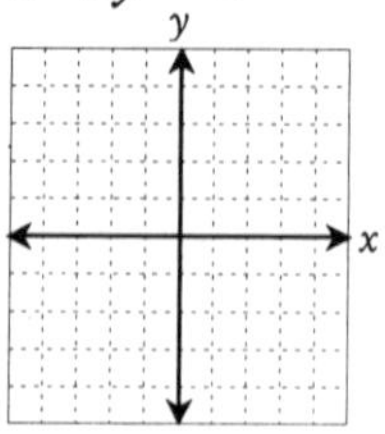

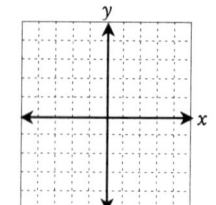

Equation:
_____   _____   _____   _____

F) Tell whether the following are linear.

47) $y = x^2 + 3$ 48) $y = -2x^2 + 3x$ 49) $y = 5 - 3x$

50) $y = \dfrac{4}{x}$ 51) $y = -5x$ 52) $y = x^3$

Answers: yes or no

47) _____   50) _____
48) _____   51) _____
49) _____   52) _____

G) Application problems

53) The graph shows the cost of renting a truck based on the number of miles traveled.

   a) What is the basic fee for renting a truck? _____

   b) What is the total cost of renting a truck and driving it 200 miles? _____

   c) If the total cost of renting a truck is $125, how many miles were traveled? _____

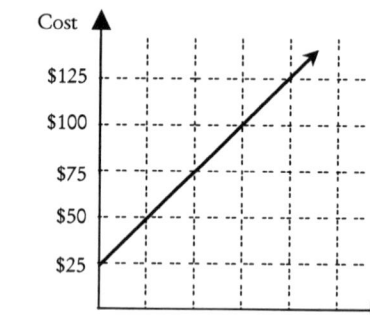

54) The attendance at girls' basketball games over the last five years is shown on the graph. The gym holds 350 fans.

   a) What is the slope of the line? _____

   b) What does the slope represent? _____

   c) If attendance continues to increase at the same rate, during which year will the gym be filled to capacity? _____

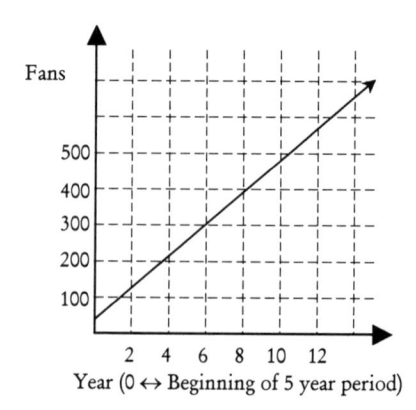

Meeting the Challenge...Mathematics

TEST PRACTICE

MULTIPLE–CHOICE QUESTIONS: Write the letter of the correct answer.

1) The slope of the line containing points (1, 4) and (–3, 1) is

   a. $\frac{3}{2}$ b. $-\frac{1}{2}$ c. $\frac{3}{4}$ d. –1

2) Which expression is a linear equation?

   a. $x^3 = 2x$ b. $x^2 + y^2 = 9$ c. $y = 3x + 2$ d. $y = \frac{5}{x}$

3) The y-intercept of the graph of $y = 4x - 3$ is

   a. 4   b. –4   c. 3   d. –3

4) Which is the equation of a line with $m = \frac{5}{2}$ and y-intercept = 3 ?

   a. $y = \frac{5}{2}x + 3$ b. $y = \frac{5}{2}x - 3$ c. $y = -\frac{5}{2}x + 3$ d. $y = -\frac{5}{2}x - 3$

5) The equation shown on the graph is

   a. $y = -3x - 2$ b. $y = 3x - 2$

   c. $y = \frac{1}{3}x - 2$ d. $y = -\frac{1}{3}x - 2)$

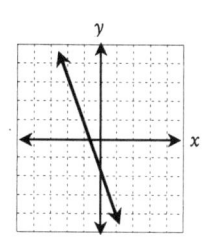

SHORT ANSWER QUESTIONS:

6) Give examples of positive, negative, zero and no slope if you were a professional mountain climber.

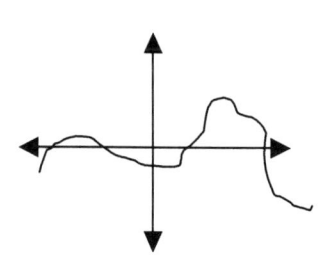

   a) positive slope:
   _____

   b) negative slope:
   _____

   c) zero slope:
   _____

   d) no slope:
   _____

7)

   a) How many zeros does this function have? ____

   b) Mark the zeros on the graph.

   c) Tell how to determine the zeros of this function.
   _____
   _____

© 2003 Orange Frazer Press. All Rights Reserved.

Meeting the Challenge...Mathematics

EXTENDED RESPONSE QUESTIONS:

8) Steve delivers pizza five nights a week. He earns $25 per night and averages a $3 tip per delivery.

   a) Develop an equation to find Steve's earnings (y) each night based on the number of deliveries (d).

   Equation: _____

   b) Complete the table.

Number of deliveries (d)	5	10	15	20	25
Steve's daily earnings (y)					

   c) Plot the data from the table in part b onto the graph.

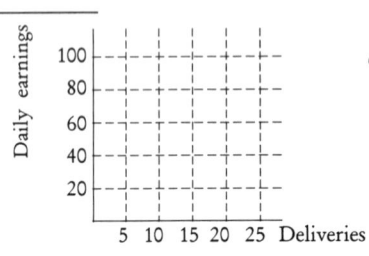

   d) What would Steve's earnings be if he made no deliveries? _____

9) Write an equation for the line segment on the graph.

   a) Slope = _____ y-intercept = _____ Equation: _____

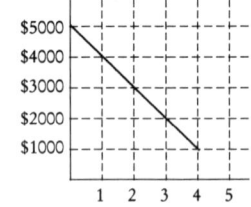

   b) Describe a real world situation that could be represented by this graph.
   _____

   c) Within the context of your real world situation, explain the meaning of the slope and the y-intercept.

   Slope: _____ y-intercept: _____

10) Given: slope = 2   A = (–2, –5)

    a) Before sketching the line, describe how to find additional points on the line by using the slope and the given point.
    _____
    _____

    b) Find three additional points (B, C, and D) on the line by using your method.

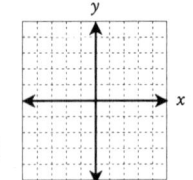

    B = (,) C = (,) D = (,)

    c) Verify that points A, B, C, and D are linear by graphing them.

CLOSURE

Let's return to the focus problem on page 129. We'll develop a formula and a graph to determine the price of each choir member's t-shirt based on the number of letters in each name.

1) The total price of the t-shirt is the fixed cost of $8 plus $0.50 for each letter in the name. Let x = the number of letters in the name. Therefore, $0.50x$ = the cost of the lettering. The **formula** to find the total price (P) of the t-shirt is $P = 0.5x + 8$.
2) The price for Charlene's t-shirt: $P = 8 + (0.5)(8) = 8 + 4.00$ **Answer:** $P = \$12.00$
3) Sketch the graph.
4) The price of Charlene's t-shirt is point A (8, 12).
5) Yes, $12.00.
6) The slope of the graph is 0.5 meaning the price of a t-shirt increases $0.50 for each letter.
7) The y-intercept is $8, which means that the basic price of the t-shirt with no lettering is $8.

Meeting the Challenge...Mathematics

MATHEMATICAL PROCESSES
Finding Answers in Different Ways

PURPOSE: In this lesson you will recognize and use equivalent representations and related procedures for a mathematical concept.

A) The store manager that supervises Tamika at her after school job requires her to perform math related transactions quickly and accurately. These transactions include: making change in several ways, calculating the sale price at the register and figuring sales tax.

The cash register tells the clerk exactly how much change to give back to the customer. For a special sale day, several additional check-out stations are set up to handle the increase in sales. Tamika will be working at one of the special check-out stations with a small adding machine instead of a cash register. The following problems will give you an idea of the calculations she will need to do without the help of the cash register.

1) Based on a total sale of $56.18, tell how Tamika would give change (coins and bills) for a $100 bill without using subtraction.

2) Discuss another way change could be given to the customer. Determine the change to be given and verify that both methods (problems #1 and #2) give the same result.

3) What is your county's present sales tax rate? _____

   a) Based on this current rate, find the sales tax for $38.15 in sales. _____

   b) Give the amount of the total bill. _____

4) In problem 3, we solved for the amount of sales tax and added it to the amount of sales to get the total bill.

   a) Give a combined percentage that would include both the original price (100%) and the sales tax (see problem #3). Explain your reasoning.

   b) This single percentage simplifies the process and allows you to use one step to find the total bill. Verify your method by using it to find the total bill in problem 3.

5) A discount can be expressed as a percentage deducted from the original price, or as a percentage of the original price. A $65 pair of jogging shoes is on sale for 15% off. Use both methods to find the sale price.

B) Refer to the graph to answer the following questions.

1) Name and give two formulas and/or methods that can be used to find the distance between points A and B on the graph.

      Name of Formula   Formula

  a) _____   _____

  b) _____   _____

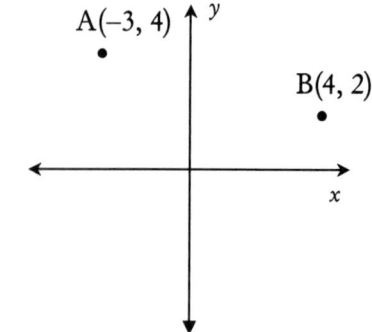

A(−3, 4)

B(4, 2)

2) Find the distance using the answer from problem 1a.

3) Find the distance using the answer from problem 1b.

C) To save money on groceries, Mrs. Watson is considering joining a discount grocery club. The Midtown Discount Grocery offers two membership plans. Plan A offers full shopping privileges for a $45 yearly fee. Plan B also offers full shopping benefits plus a 5% discount on purchases (excluding alcoholic beverages, cigarettes and cosmetics) for $175 per year. Looking at her checkbook from the previous year, Mrs. Watson found that she had spent $3628 on groceries that would have qualified for the discount.

1) Discuss the important facts that should affect her decision.

2) Which plan should she choose? Justify your choice.

3) Explain how to find the minimum amount of qualified grocery purchases that would make Plan B her better choice.

4) Solve for the answer to #3.

Meeting the Challenge...Mathematics

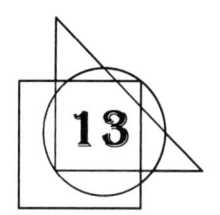

TABLES, CHARTS AND GRAPHS

Create, interpret and analyze data through
the use of tables, charts and graphs.

FOCUS: A survey was given to tenth graders at King High School. They were asked if they worked after school and, if so, to tell what type of job they held. The results are shown on the circle graph.

Refer to the graph to answer the following questions:

1. What type of job is held by the largest number of tenth graders?
2. Students that do not work represent approximately what percent of the 10th grade class?
3. Do you think this school district is in a rural area? Explain.
4. Do you think this school district is in a commercial area? Explain.
5. Name the three most popular job areas.
6. Out of 200 tenth graders, about how many would you say are not employed after school?

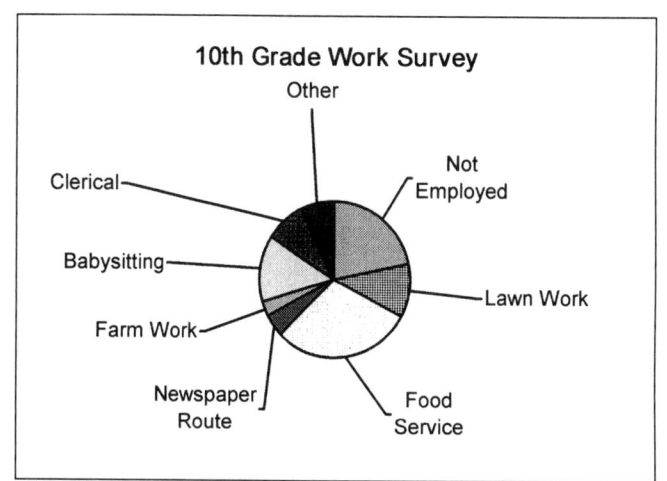

10th Grade Work Survey

Categories: Other, Not Employed, Lawn Work, Food Service, Newspaper Route, Farm Work, Babysitting, Clerical

PURPOSE: At the end of this lesson, you will be able to interpret and analyze data from line graphs, bar graphs, circle graphs, histograms, stem-and-leaf plots, box-and-whisker plots, scatter plots, pictographs, etc. You will also be able to identify patterns and trends, draw conclusions, select appropriate types of graphs, determine suitable scales, and create appropriate displays of any given data.

WHAT YOU NEED TO KNOW

MATHEMATICS AND PROBLEM SOLVING VOCABULARY:

1. Bar graph
2. Box-and-whisker plot
3. Circle graph
4. Histogram
5. Line graph
6. Median
7. Pictograph
8. Quartile
9. Scatter plot
10. Stem-and-leaf plot
11. Trend line

MEMORIZE THESE FACTS:

Summary of various graphs
1. A **bar graph** compares quantities by using a solid bar on a horizontal or vertical axis.
2. A **box-and-whisker plot** shows how data is clustered by using the median, quartiles and the extremes (smallest and largest). It does not show all the specific values in a collection of data.

© 2003 Orange Frazer Press. All Rights Reserved.

3. A **circle graph** compares quantities as parts of a whole. The whole amount is represented by the entire circle of 360°, or 100%. Each part of the circle is a percentage of the whole. A semicircle (half a circle) is 180°, or 50% of the whole. A quarter circle is 90°, or 25% of the whole.
4. A **histogram** compares data showing how it falls into different intervals within the range.
5. A **line graph** shows change and the direction of change over a period of time.
6. A **line plot** shows the spread of the data.
7. A **pictograph** compares data (much the same as a bar graph), but uses pictures and symbols.
8. A **scatter plot** allows you to analyze two sets of data at the same time by plotting corresponding numbers as ordered pairs on a graph. There might be a strong relationship between the two sets of data if the plotted points tend to line up. A trend line or 'best fit' line comes close to connecting the points on the scatter plot and is used to make predictions. *(Lesson #12)*
9. A **stem-and-leaf plot** organizes and displays every piece of numerical data in the set. The largest place value of the data is used for the stem and appears vertically. The next greatest place value is used for the leaves and appears horizontally.

BE ABLE TO PERFORM THESE OPERATIONS:

Choosing the scale for a graph

1. **Choose a simple scale that starts with zero and increases by equal intervals to the largest number.**
   Example: Graph the temperatures for the month. The coldest temperature was 22°F and the warmest was 52°F.
   Solution: Find the range of the data by subtracting the smallest value (22°) from the largest value (52°). Divide the range (30°) by the number of intervals desired. Choosing a small interval will make a tall, skinny graph. As the size of an interval increases, the graph becomes shorter and wider.
   Answer: 6 intervals with 5° in each interval

2. Sometimes starting a scale at zero is not practical. You can use a jagged line (⌇) to break the scale and show that you are not plotting values within a certain part of the scale. See page 151.

Creating graphs that show changes in data

Line graphs show how data changes over time.

Example: This line graph shows Jesse's math test scores so far this quarter.

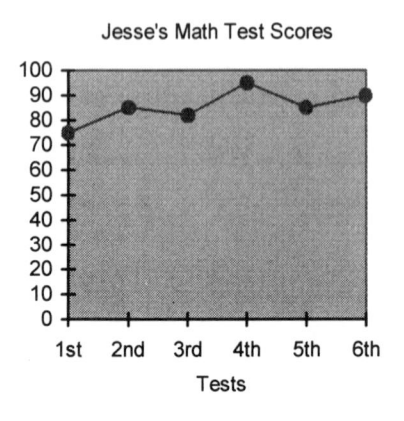

Question 1: On which test did Jesse score highest?
Answer: the 4th test

Question 2: Did Jesse's score increase or decrease from the 1st test to the 4th test?
Answer: increased

Question 3: What is Jesse's approximate grade right now?
Answer: from 85% to 87%

Question 4: Since the beginning of the quarter, has Jesse's overall score increased or decreased?
Answer: increased

Meeting the Challenge...Mathematics

Creating graphs that compare data

Typically **bar graphs, pictographs,** and **circle graphs** compare data. They may show:
- the same type of data at different times or places
- different types of data at the same time or place
- data that makes up 100% of the group

Example: During the month of August, Joe's Used Cars sold 12 minivans, 10 trucks, 2 RVs and 38 automobiles. Display the monthly sales in a bar graph, a pictograph, and a circle graph.

Bar Graph
Joe's August Sales (bar graph showing Minivans 12, Trucks 10, RVs 2, Autos 38)

Circle Graph
Joe's August Sales (circle graph with Minivans, Trucks, RVs, Autos)

Pictograph
Joe's August Sales

Minivans	🚗🚗🚗
Trucks	🚗🚗🚗
RVs	🚗
Autos	🚗🚗🚗🚗🚗 🚗🚗🚗🚗🚗

🚗 = 4 vehicles sold

Questions:
1) Which graph shows data comparisons quickly, accurately and clearly? _____
2) Although easy to read, the _____ is a less accurate graph due to the need to round to a convenient unit.
3) Use a _____ to show relationships of parts of data to each other and to the whole.
4) According to the pictograph, Joe sold _____ trucks in August.
5) According to the circle graph, minivans and trucks make up what fraction of Joe's August sales? ____
   Answers: 1) bar graph   2) pictograph   3) circle graph   4) 10   5) $\frac{1}{3}$

Creating graphs that show how data is grouped

Box-and-whisker plots, line plots, stem-and-leaf plots, and **scatter plots** order, group and display data.

Example 1: An automated traffic counter recorded the number of vehicles traveling through an intersection each morning. The data for 15 days was reported as follows:
37, 91, 20, 25, 88, 76, 51, 91, 25, 29, 67, 78, 98, 45, 30
a) Organize the data using a stem-and-leaf plot. The tens are the stems and the ones are the leaves.
b) Construct a box-and-whisker plot to clearly display the distribution of the data across the range.

a) Stem-and-Leaf Plot   Ordered Data

Stem	Leaf
9	1 1 8
8	8
7	6 8
6	7
5	1
4	5
3	7 0
2	0 5 5 9

98, 91, 91
88
78, 76
67
51
45
37, 30
29, 25, 25, 20

b) Box-and-Whisker Plot
Using the ordered data from the stem-and-leaf plot, the median is the middle number which is 51. The 1st quartile is the median of the lower half which is 29. The 2nd quartile is the same as the median which is 51. The 3rd quartile is the median of the upper half which is 88.

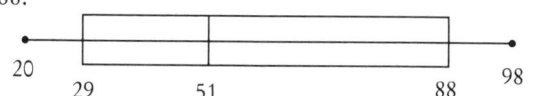

(box-and-whisker plot: 20, 29, 51, 88, 98)

Observation: Since the median is below the middle of the range, the data is clustered in the lower half of the range.

Example 2: Twenty students from Mr. Walker's senior homeroom were asked to keep records of the number of hours worked per week. Use a line plot and a histogram to organize this data. The hours reported for last week were:
9, 16, 4, 15, 10, 12, 10, 9, 18, 10, 12, 18, 4, 8, 10, 15, 10, 4, 9, 12

Solution: Once the line plot is completed, the histogram is organized into four intervals. The total x's within each interval on the line plot are shown on the histogram. Looking at both graphs, you can see the largest cluster is in the 6-10 interval.

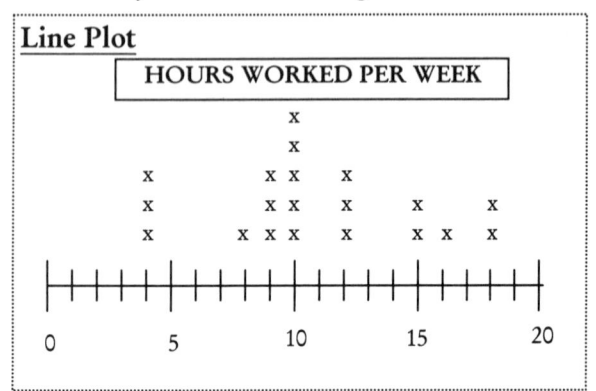

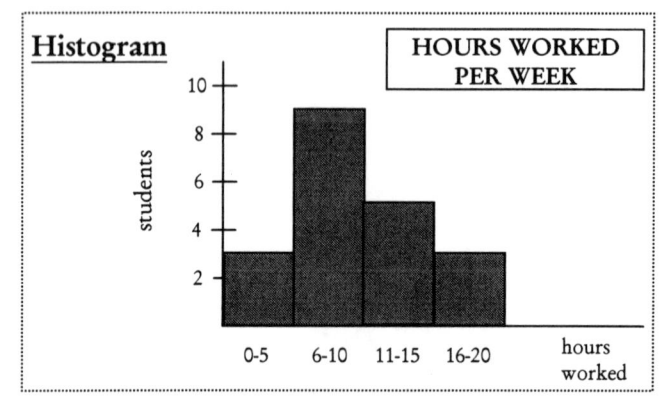

BE ABLE TO PERFORM THESE CALCULATOR OPERATIONS:

Use your calculator as needed when working with numerical data.

GUIDED PRACTICE

A) Choose the most appropriate graph to use to display the following data.

    a) circle graph   b) line graph   c) histogram   d) bar graph

1) the changes in the price of a stock during the year _____

2) the number of birds that appear at the feeder during 3-hour intervals _____

3) a budget showing how you spend your monthly income _____

4) the number sold of the three most popular video games _____

B) Based on the given data, select an appropriate scale for a graph.

    Tell: a) the range   b) the size of the interval   c) how many intervals to use on the graph

5) data ranging from 0 to 900 units sold

    a) _____   b) _____   c) _____

6) data ranging from $15,000 to $120,000

    a) _____   b) _____   c) _____

7) data ranging from −9°F to 38°F

    a) _____   b) _____   c) _____

8) data ranging from 2,500,000 to 4,000,000 people

    a) _____   b) _____   c) _____

Meeting the Challenge...Mathematics

C) **Using the given data, construct the graph.**

9) A consumer research company surveyed the total cost of 25 food items in 6 Ohio cities. The results are shown on the table.

City	Cost
Anna	$146
Cleveland	$175
Port Clinton	$201
Steubenville	$128
Toledo	$165
Waverly	$155

Make a bar graph for the table. Title the graph, label both axes and select a suitable scale for the data. Draw a bar to represent each city.

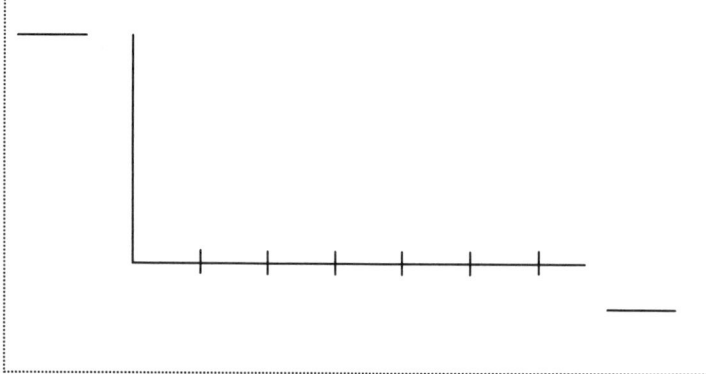

What scale would you use to display this data? _____

The data in the table is quite varied. Explain. _____

10) Could a pictograph be used to display this data? _____ Explain. _____

11) Could a circle graph be used to display this data? _____ Explain. _____

12) Make a line graph for the data in the table.

a) The title of the graph is _____.

b) The label for the vertical axis is _____.

c) The label for the horizontal axis is _____.

d) What is the range of temperatures? _____

e) How many intervals should be used? _____

f) What is the size of each interval? _____

g) Give the values for each interval.
_____

Monthly Low Temperatures	
July	61°
August	58°
September	45°
October	31°
November	28°
December	12°

© 2003 Orange Frazer Press. All Rights Reserved.

The test scores for Mrs. Jaxon's math class are shown.

Scores: 95, 75, 61, 81, 88, 90, 85, 79, 99, 87, 95, 85, 88, 93, 76, 68, 99, 95, 88, 90, 65, 88

13) Organize these scores in a stem-and-leaf plot.

Stem	Leaf

Refer to the stem-and-leaf plot to answer these questions.

a) What is the range of the scores? _____

b) Which score occurred most often? _____

c) Which stem had the most scores? _____

14) Use the information from the stem-and-leaf plot to construct a box-and-whisker plot.

Refer to the box and whisker plot to answer these questions.

a) The extreme scores are _____ and _____.

b) The median score is _____.

c) The lower quartile score is _____.

d) The upper quartile score is _____.

15) Organize these same scores into a histogram.

Refer to the histogram to answer these questions.

a) How many intervals should be used? _____

b) What is the range of each interval? _____

c) Which interval contains the most scores? _____

16) Every day Denise usually spends 7 hours in school, 2 hours doing homework, 4 hours working at a part-time job, 8 hours sleeping and 3 hours doing miscellaneous activities (having fun, eating, etc.). Make a circle graph to show how Denise spends an average school day.

a) Set up ratios to find the percents that represent each portion of her day.

School time _____ = _____

Homework time _____ = _____

Job time _____ = _____

Sleep time _____ = _____

Miscellaneous time _____ = _____

b) Use the percentages to make a circle graph. Estimate the size of each section of the circle and label each section.

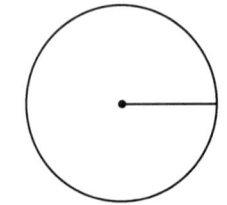

Meeting the Challenge...Mathematics

INDEPENDENT PRACTICE

A) Which graph (bar graph, line graph, histogram, circle graph, stem-and-leaf plot, box-and-whisker plot, or pictograph) would you use in the following situations?

1) If you want to show the number of occurrences in an interval, use a _____.
2) If you want to show the relationships of the parts to the whole, use a _____.
3) If you want to organize data from an unordered list, use a _____.
4) If you want to show the distribution of data using the largest and smallest values and the distribution of the data with respect to the median and quartiles, use a _____.
5) If you want to show how temperatures change throughout the month, use a _____.
6) If you want to show a comparison of data and be very accurate with values use a _____.
7) If you want to show a relative comparison of data in an attractive display, use a _____.
8) If you want to show the changes in the interest rate, use a _____.

B) Use each graph to answer the questions.

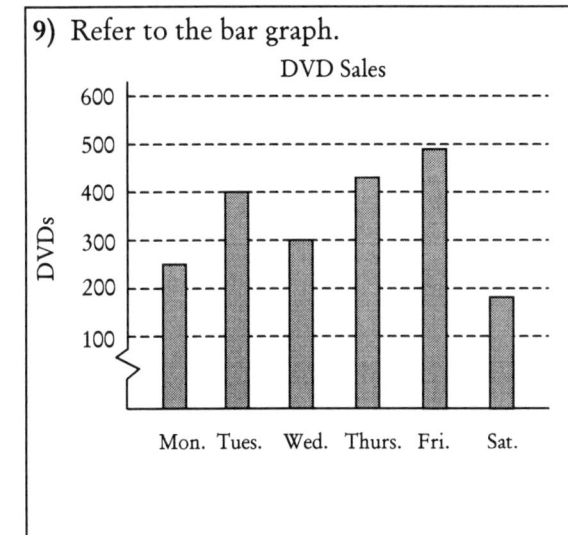

9) Refer to the bar graph.

   DVD Sales (bar graph showing DVDs sold Mon–Sat: Mon 250, Tues 400, Wed 300, Thurs 430, Fri 490, Sat 190)

   a) Sales for Friday are more than what two days combined? _____ and _____
   b) The manager would like to see Wednesday's sales increase by 10%. How many more DVDs need to be sold? _____
   c) The store is open later on Tuesday, Thursday and Friday. Would you say this is a good idea? _____ Explain. _____
   d) Tuesday's sales represent approximately what percent of the total sales of DVDs for the week? _____
   e) There are six sales employees working on Friday. Based on sales, about how many sales employees should be working on Saturday? _____ Why? _____

10) This graph shows the price of XYZ stock at two-month intervals during one year.

    Cost of XYZ Company Stock (line graph, Feb–Dec, $14–$22)

    a) Complete this line graph.
    b) The price of the stock at the end of the year was $_____.
    c) The greatest gain in price occurred during the two months of _____ and _____.
    d) The price stayed the same during _____ and _____.
    e) Between which two time periods was there a 20% increase in the stock price? _____ and _____
    f) What does the ⌇ in the graph mean? _____

© 2003 Orange Frazer Press. All Rights Reserved.

Meeting the Challenge...Mathematics

The table and the circle graph display a breakdown of the expenses for maintaining a car.

Expense Item	Cost
Gas/oil	28%
Repairs and maintenance	15%
Insurance	37%
Taxes	10%
Parking fees/tolls	3%
Miscellaneous	7%

Costs of Owning a Car

11) Tell which item is represented by each letter in the circle graph.

a) _____   d) _____

b) _____   e) _____

c) _____   f) _____

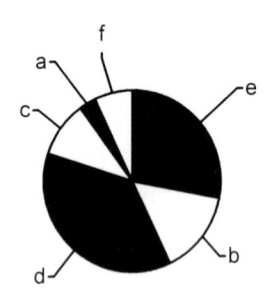

12) Erik allows $130 a month to maintain his car. Based on the table and circle graph, how much money is allocated for gas and oil?

Answer: _____

13) If Erik needs to allocate $80 a month for insurance, how much should be budgeted for auto maintenance each month?

Answer: _____

Use the scatter plot graph showing sales of cosmetics for the first six months of the year to answer the questions.

Cosmetic Sales

14) A 'best fit' trend line can be drawn on this scatter plot. Why is this true?

15) Draw a trend line and project September's sales. $_____

16) Explain when a trend line cannot be drawn.

17) If the data does not line up well enough for a trend line, give some ideas of how a manager could project sales.

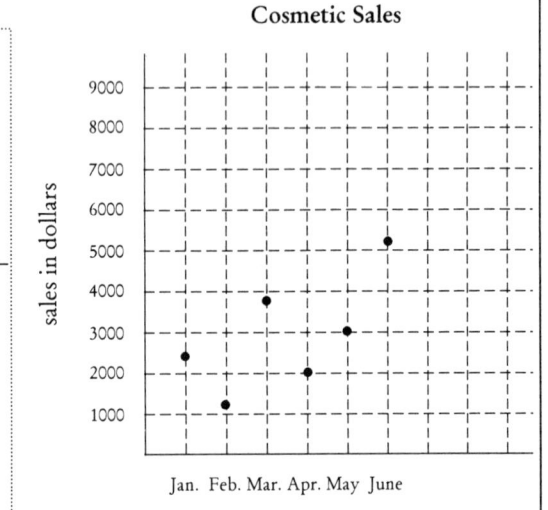

Meeting the Challenge...Mathematics

A gas mileage survey was given. These are the results in miles per gallon.
25, 29, 38, 32, 25, 31, 32, 29, 18, 21, 23, 44, 38, 19, 17, 20, 32, 28, 24, 18, 35, 17

18) Organize the data using a stem-and-leaf plot.

Stem-and-Leaf Plot	Ordered Data
Stem | Leaf	

19) Make a histogram to show the results of a gas mileage study. Use the following interval ranges: 17-21, 22-26, 27-31, 32-36, 37-41, 42-46

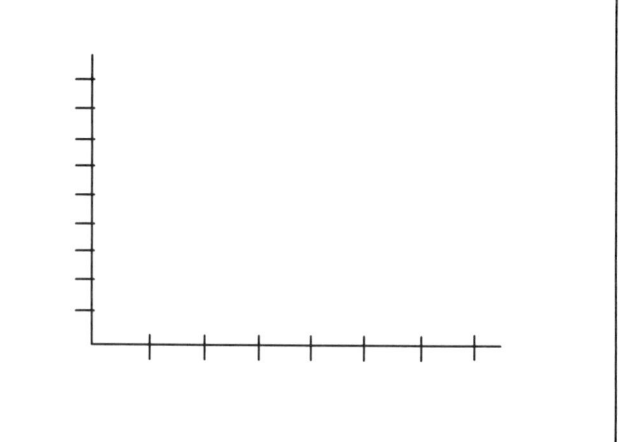

20) Use the ordered data from the stem-and-leaf plot above to find the following:

   a) median _____

   b) lower quartile _____

   c) upper quartile _____

   d) extremes _____ and _____

   e) Construct a box-and-whisker plot.

TEST PRACTICE

MULTIPLE-CHOICE QUESTIONS:

1) A line graph _____

    a. organizes data by intervals.   b. shows data changes over time.

    c. compares data.   d. shows how data is clustered.

2) The 'plot' graphs such as line plots and stem-and-leaf plots _____

    a. organize data.   b. use pictures to display data.

c. use bars to display data.   d. show data changes over time.

3) To show the changes in a stock price over the past year, use a   _____

    a. bar graph   b. histogram

    c. line graph   d. box-and-whisker plot

4) Values on a bar graph range from $250 to $1000. If ten intervals are used starting with $250,   _____
the size of each interval is

    a. $100   b. $50
    c. $75   d. $125

5) Use the circle graph to tell which class donated $\frac{1}{3}$ of the money needed for the spring play.   _____

    a. 9th
    b. 10th
    c. 11th
    d. 12th

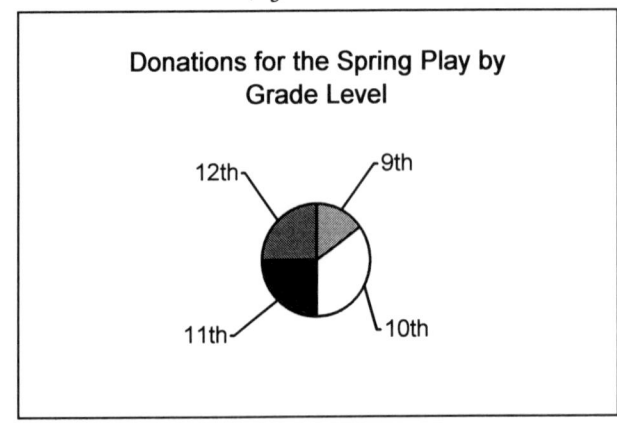

Donations for the Spring Play by Grade Level

SHORT ANSWER QUESTIONS:

6) Use a stem-and-leaf plot to organize the test data.
75, 66, 58, 72, 69, 79, 88, 92, 61, 82, 50, 99, 85, 90, 87, 79, 60, 72, 88, 95, 83

Stem | Leaf

7) Display the organized data in a histogram. Use groups of ten for each interval.

Meeting the Challenge...Mathematics

EXTENDED RESPONSE QUESTIONS:

8) A teacher kept track of the number of hours he exercised each month and the weight that he lost. Plot the data on the graph.

Exercise time (in hours)	0	5	10	15	20	25
Weight loss (pounds)	1	3	2	4	5	8

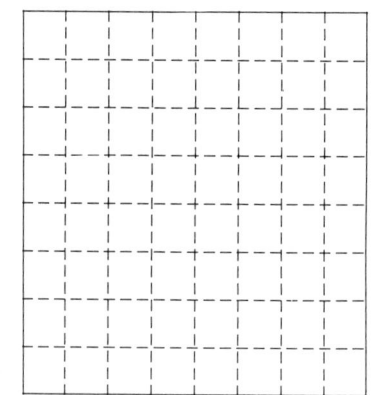

a) Draw the 'best fit' trend line.

b) Based on the trend line, how much weight could the teacher expect to lose if he exercises 40 hours a month? ___

9) Refer to the box-and-whisker plot below to answer the following questions.

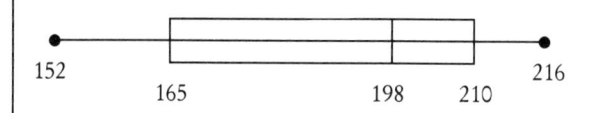

152   165   198   210   216

a) The median is _____.

b) The lower extreme is _____.

c) The upper extreme is _____.

d) The lower quartile is _____.

e) The upper quartile is _____.

f) Describe how the data in your collection is distributed.
_____
_____

Now create a data collection of 11 numbers that could be represented by the same box-and-whisker plot.
_____

Verify that the answers from your data above match the answers from the given box-and-whisker plot.

a) The median is _____.

b) The lower extreme is _____.

c) The upper extreme is _____.

d) The lower quartile is _____.

e) The upper quartile is _____.

f) Describe how the data in your collection is distributed.
_____
_____

10) This line graph shows Tanya's typing speed (in words per minute) on 8 tests.

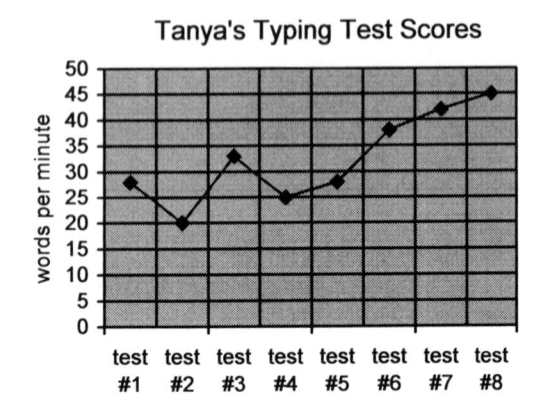

Tanya's Typing Test Scores

Question 1: On which typing test did Tanya score the lowest?
Answer: _____

Question 2: Did Tanya's score increase or decrease from the 1st test to the 4th test?
Answer: _____

Question 3: Should you expect Tanya to type at least 30 words per minute?
Answer: _____

Question 4: Since the beginning of the quarter, has Tanya's overall score increased or decreased?
Answer: _____

Question 5: Between which two tests was there the most improvement? _____ and _____

Question 6: How is this graph helpful?
Answer: _____
_____

CLOSURE

Let's return to the focus problem on page 145.

1) food service
2) close to 25%
3) Based on the small number of students doing farm work and the large number of students working in food service, this school district is not in a rural area.
4) It could be that this school district is in a commercial area based on the large number of students working in food service, but there is not enough information to state this as a fact.
5) food service, babysitting, lawn work
6) Use 25% as the percentage of students not working.   Answer: 50 students (200 × 25% = 50)

Meeting the Challenge...Mathematics

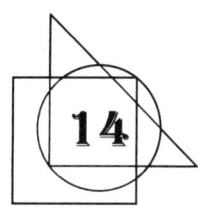

MEASURES OF CENTRAL TENDENCY

Choose and apply measures of central tendency (mean, median, and mode) and variability (range and visual displays of information).

FOCUS: Let's use our mathematics skills to determine whether a home is affordable.

Mrs. Hunt is taking a new job and must relocate to a new community. She can afford a house in the $65,000 to $75,000 price range. In the newspaper, she saw a listing for five homes that sold recently in the neighborhood where she is looking. Their sale prices were $82,000; $60,500; $73,500; $65,300; and $76,100. Can Mrs. Hunt afford a house in this neighborhood?

PURPOSE: When you have completed this lesson, you will be able to choose, find and apply measures of central tendency (mean, median and mode) and variability (range and deviation) to learn more about a set of data.

WHAT YOU NEED TO KNOW

MATHEMATICS AND PROBLEM SOLVING VOCABULARY:

1. Average
2. Average deviation
3. Mean
4. Measures of central tendency
5. Median
6. Mode
7. Outlier
8. Quartile
9. Range
10. Standard deviation
11. Variability

MEMORIZE THESE FACTS:

1. The three measures of **central tendency** are used to describe what is typical about a set of data.
   - The **mean** is found by adding the items, then dividing by the number of items that were added.
   - The **median** is the middle number in an ordered set of data. If there is an even number of data items, there are two middle numbers. The median is the mean of these two numbers. The median splits the ordered data into two parts.
   - The **mode** is the value or item that occurs most often in a data set.
2. The **range** is a measure of the spread of the data. Find the range by subtracting the largest and smallest numbers in the data set.
3. The **quartile** divides an ordered set of data into four equal parts.
4. The **deviation from the mean** is the difference between each item of data and the mean.
5. **Average deviation** tells us how the numbers in the data set differ from the mean. A lower average deviation tells us that the data tends to be clustered closer to the mean. A higher average deviation tells us that the data is scattered farther from the mean.
6. When describing data using measures of central tendency and the range, keep these ideas in mind.
   - Reporting both the mean and the range of a set of data gives a better summary of the data than reporting just the mean.
   - An item that is far away from the rest of the data is called an outlier. The range can be dramatically impacted by an outlier and should be considered when interpreting data.
   - When the mean or the median does not seem to accurately represent the data, look at another measure of central tendency.

© 2003 Orange Frazer Press. All Rights Reserved.

BE ABLE TO PERFORM THESE OPERATIONS:

Choosing the mean, median, or mode

Depending on the nature of the data, one measure of central tendency may be better than another to represent a 'typical' number in the set.

Case 1: If a set does not have an extreme high or low, explore using the **mean**. If the data items are closely grouped, the mean will be close to the data and be a good 'typical' value. If there is an extreme high or low data item, the mean will not be as close to the main cluster of data and will not accurately describe the data.

Example 1: In six basketball games, Jan scored 12, 16, 15, 16, 12, and 15 points. Find his average score per game. Solution: Since the scores are closely grouped, use the mean. Answer: mean = 86 ÷ 6 = 14.3 points *Now, let's change one score in this example so the data set is not as closely grouped. Look at Example 2.*

Example 2: In six basketball games, Jan scored 12, 16, 15, **2**, 12 and 15 points. Find his average score per game. Solution: In this example, there is an extreme low value of 2. The mean is now equal to 12 (72 ÷ 6). The mean is lower, meaning that the extreme low score (2) brought down the three higher scores (15, 15, and 16). The mean no longer is a good 'typical' score to describe this data. Now look at case 2.

Case 2: If a set has extreme highs or lows and does not have big gaps in the middle, explore using the **median** as a 'typical' value to describe the set.

Example: Look back at Example 2. Jan's two point game reduced the mean of the set significantly and made it a less desirable way to describe the data. Solution: Because of the extreme score, the median is a better way to describe the data. First, organize the data set in order from least to greatest. Scores: 2, 12, 12, 15, 15, 16 Answer: The median score is halfway between 12 and 15 or 13.5 [(12 + 15) ÷ 2] making it a better 'typical' value.

Case 3: If a set has repeating data numbers, the **mode** (the value or item that occurs most often) may be used to describe the data. There may be more than one mode in a set.

Example: Maxwell High School sold alumni jackets in Small, Medium, Large and X-Large sizes. The following orders were taken: 5 Small, 12 Medium, 28 Large and 16 X-Large. The typical size of the jackets sold is **Large** which is the mode.

Finding missing values in a set of data

When the average or mean is given in a problem, use it to find the sum of the items in the set. Multiply the average by the number of items in the set to find the sum. Once the sum of the items is known, subtract the known items to find the missing data item.

Example: Scott averaged 11 points in the first five games of the basketball season. He scored 8 points in the first game, 9 points in the 2nd game, 6 points in the 3rd game and 15 points in the 5th game. How many points did he score in the 4th game? Solution: Scott scored 55 points in the first five games. We know that he scored 38 points (8+9+6+15) plus what he scored in the 4th game. Answer: In the 4th game, Scott scored 17 (55-38) points.

Meeting the Challenge...Mathematics

Arranging data and describing variability

Use the data in the box for these examples.

1. **Find the range of the set of ordered data.**
   Answer: 73 − 36 = 37
2. **Arrange the ordered data into quartiles. Start by finding the median.**
   Solution for the median or the 2nd quartile: There are 22 items in the data set. The median is the middle number. Since there is an even number of data items, there are two middle numbers. The median is the mean of these two numbers which is 57.
   Solution for the 1st quartile: The 1st quartile is the median of the lower half of the data which is 45.
   Solution for the 3rd quartile: The 3rd quartile is the median of the upper half of the data which is 63.
3. **Find the deviation of each value from the mean.**
   Solution: Solve for the mean.   Answer: 55
   To solve for each value's deviation from the mean, find the difference between the data value and the mean. See the box.
4. **Find the average deviation.**
   Solution: Find the total deviation from the mean (206) and divide by 22 (the number of items).
   Answer: $9.\overline{36}$

Ordered Test Scores

Deviation from mean	Data	
18	73	
17	72	
15	70	
14	69	
13	68	
8	63	← 3rd quartile
5	60	
4	59	
3	58	
2	57	
2	57	← median
2	57	
0	55	← Mean
5	50	
5	50	
7	48	
10	45	← 1st quartile
11	44	
14	41	
16	39	
16	39	
19	36	

BE ABLE TO PERFORM THESE CALCULATOR OPERATIONS:

Use your calculator as needed when working with numerical data.

GUIDED PRACTICE

A) Which measure of central tendency is described in these problems?

1) The class average on a math test is 81%.

2) The middle score on 13 typing tests is 45 words per minute.

3) Jay scored a 6 on five out of nine holes at Community Golf Course.

4) The most popular color of jacket is blue.

5) Six employees are paid more than Stacy and six are paid less.

6) Find the average price for three shirts.

Answers:
1) _____
2) _____
3) _____
4) _____
5) _____
6) _____

B) **Solve for the mean and the median. Tell if one measure gives a better description of the data than the other.**

7) During a 5-hour shift, the manager at Ace Car Wash tallied the number of vehicles washed each hour. 1st hour → 5 vehicles, 2nd hour → 8 vehicles, 3rd hour → 30 vehicles, 4th hour → 12 vehicles, 5th hour → 7 vehicles

   Organize the data and answer the questions.

   a) Organize the data. _____   mean = _____   median = _____

   b) Which measure gives a better description of the data? _____

   c) Why? _____

8) Last week in Mr. Keller's homeroom, 5 students earned $38, $56, $46, $42 and $50. Organize the data and answer the questions.

   a) Organize the data. _____   mean = _____   median = _____

   b) Which measure gives a better description of the data? _____

   c) Why? _____

C) **Apply measures of central tendency and variability to this problem.**
Mr. Harris divided his class into two groups. He recorded the grades and the attendance for both groups. He wanted to know which group had more variation in their grades. To answer this question, he taught his class to find the mean, the deviation from the mean and the average deviation of their midterm averages. Answer the questions using the table that shows midterm averages for each group.

Student	1	2	3	4	5	6	7	8	9
Browns	79%	69%	88%	84%	97%	95%	90%	83%	89%
Bengals	75%	84%	79%	86%	91%	88%	77%	95%	83%

9) Find the mean of each group.

   a) mean of the Browns: _____

   b) mean of the Bengals: _____

10) Give the range of each group's scores.

   a) range of the Browns: _____

   b) range of the Bengals: _____

11) Find the deviation from the mean for each member of both groups.

Student	1	2	3	4	5	6	7	8	9
Browns									
Bengals									

12) Find the average deviation.

   a) Browns: _____

   b) Bengals: _____

13) Compare the average deviations.

   a) Which team differs more from the mean? _____

   b) What observations can you make about the midterm averages of the two teams?

Meeting the Challenge...Mathematics

D) **Solve these application problems.**

14) On the first three days of a four-day trip, Tamika's family drove 268 miles, 320 miles, and 290 miles.

   a) How far is their destination if they averaged 300 miles a day? _____

   b) How far did they travel the fourth day? _____

15) **Riley** researched the price of gasoline. Every week she filled the family car and recorded the price per gallon. The prices for 8 weeks were:

   $1.48, $1.36, $1.42, $1.16, $1.28, $1.22, $1.36, $1.58

   a) the mean = _____

   b) What are some possible reasons for the prices changing so much week to week?

INDEPENDENT PRACTICE

A) **Solve these word problems.**

1) The mean of the weights of three friends is 108 pounds. How much does the 4th friend weigh if the average weight of all four friends is 115 pounds?

   _____ pounds

2) Together three friends weigh 380 pounds. How much does the 4th friend weigh if the average weight of all four friends is 125 pounds?

   _____ pounds

3) Replace one number in the list so that the mean of the new list is 12.

   Old list: 8, 4, 11, 17

   New list: _____

   Verify the mean of the new list.

4) Replace one number in the following list so that the median of the new list is 15.

   Old list: 8, 12, 13, 16, 19

   New list: _____

   Verify the median of the new list.

B) **Organize the data and solve for the mean and the median. Tell which measure gives a better description of the data.**

5) The high temperatures for a week in April were 68°F, 55°F, 63°F, 51°F, 48°F, 32°F, and 46°F.

   a) Organize the data.

   b) the mean = _____   c) the median = _____

   d) Which measure gives the better description of the data? _____
   Explain.

6) Four used cars sold for $2900, $2600, $8200 and $3200.

   a) Organize the data.

   b) the mean = _____   c) the median = _____

   d) Which measure gives the better description of the data? _____
   Explain.

7) The number of hours that ten students spent on the internet during one week is as follows:
75, 12, 25, 16, 50, 38, 65, 58, 8, 62

a) Organize the data.

b) the mean = _____

c) the median = _____

d) Which measure gives the better description of the data? _____
Explain. _____
_____

8) The drivers for the Hurry Up Delivery Service must report the amount of time they spend at each stop. The minutes spent for Monday were reported as follows: 3, 1, 6, 10, 10, 9, 12, 3, 1, 1, 5

a) Organize the data.

b) the mean = _____

c) the median = _____

d) Which measure gives the better description of the data? _____
Explain. _____
_____

9) In the last 10 basketball games, Brooklyn scored 20, 12, 0, 15, 25, 18, 20, 24, 30, and 28 points.

a) Organize the data.

b) the mean = _____

c) the median = _____

d) Which measure gives the better description of the data? _____
Explain. _____
_____

10) Your grades on four tests were 35, 86, 92, and 93.

a) Would the mean or the median be the fairer measure of central tendency to use? _____

b) Show your calculations, work and explanation.

C) Apply measures of central tendency and variability to the following problem.

The FlySafe Airlines and the FlyOnTime Airlines reported the amount of time that their flights were delayed each day.

Airline	Flight 1	Flight 2	Flight 3	Flight 4	Flight 5	Flight 6	Flight 7	Flight 8	Flight 9
FlySafe	25	15	6	0	10	14	20	3	0
FlyOnTime	3	0	30	9	12	19	30	0	11

11) Find the mean of each group.

a) mean of FlySafe Airlines: _____

b) mean of FlyOnTime Airlines: _____

12) Give the range of each group's scores.

a) range of the FlySafe Airlines: _____

b) range of the FlyOnTime Airlines: _____

Meeting the Challenge...Mathematics

13) Find the deviation from the mean for each airline.

Airline	Flight 1	Flight 2	Flight 3	Flight 4	Flight 5	Flight 6	Flight 7	Flight 8	Flight 9
FlySafe									
FlyOnTime									

14) Find the average deviation.

    a) FlySafe: _____

    b) FlyOnTime: _____

15) Compare the average deviations for the two groups.

    a) Which airline differs more from the mean? _____

    b) What observations can you make about the amount of delays of the two airlines?

    _____

TEST PRACTICE

MULTIPLE-CHOICE QUESTIONS:

1) If a data set has an even number of items, you know:

    a. It has two modes.

    b. It doesn't have a mode.

    c. The median is in the upper quartile.

    d. The median is the mean of the two middle items.

2) Isaac's geometry grade was 82% after the first three tests. He scored 90% on the fourth test. What was his average after the 4th test?

    a. 86%   b. 87%   c. 85%   d. 84%

3) A lower average deviation is saying:

    a. The data is clustered around the median.

    b. The data is clustered around the mean.

    c. The data has two modes.

    d. The answer is probably not right.

4) In a data set of 11 items, where is the 9th term located?

    a. at the median   b. upper quartile   c. upper deviation   d. at the mode

5) Dorian's bowling scores for her first three games were 110, 145 and 95. What score does she need to bowl in the last game to end up with an average of 120 for all four games?

    a. 130   b. 145   c. 120   d. 160

SHORT ANSWER QUESTIONS:

6) In an electronics store, the mean salary of 11 employees is $400 per week. The manager's salary of $600 per week is included in this average.

    a) How does his salary affect the mean? _____

    b) If the other 10 employees all receive the same salary, their weekly salary is $_____.

Meeting the Challenge...Mathematics

7) Keeshon scored 78, 86, 77 and 75 on four tests. Find the least number of points she can score on her next test to have at least an 80 average.

8) Use this data set for the following problems: 5, 6, 9, 11, 14, 15

   a) Replace one of the numbers in the list so the mean of the new list of numbers is 12. Write the new list.

   b) Replace one of the numbers so the median of the new list of numbers is 11. Write the new list.

EXTENDED RESPONSE QUESTIONS:

9) The girls' basketball all-time free throw percentage record is 92% at North-Forty High School. Kathy's goal is to break that record. Up to this point, she has made 162 out of 180 free throws. There are four games left in the season. On the average, she will have 12 free throw attempts per game. Can Kathy break the record? Explain your approach to solving this problem. Support your answer with formulas, calculations and other factual data.

10) Every six months the All Sports Store offers a bonus to their best employee. Use the data, the mean and the median to make a recommendation for the best employee. Explain the reasons for your choice.

Employee	July Sales	Aug. Sales	Sept. Sales	Oct. Sales	Nov. Sales	Dec. Sales
Dylan	$6535	$10,246	$9045	$3905	$12,519	$20,654
Christian	$2249	$14,905	$6658	$5555	$12,900	$20,500

CLOSURE

Let's return to the focus problem on page 157.
The mean of $71,480 is within Mrs. Hunt's price range.
To find the median, organize the prices from least to greatest. $60,500; $65,300; $73,500; $76,100; $82,000
The median home price is $73,500, which is still in Mrs. Hunt's price range.
Mrs. Hunt should be able to find a house within her price range in the neighborhood where she is looking.

Meeting the Challenge...Mathematics

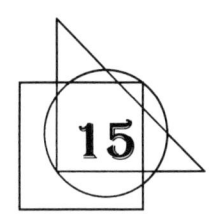

PROBABILITY

15

Represent and interpret possible outcomes and calculate probabilities.

FOCUS: Let's use mathematics to determine our number of choices.

The twelve reserve cheerleaders are buying new sweaters for the football season. Their choices are a blue or yellow sweater with white, black or green lettering and a traditional or contemporary flag logo. Is it possible for every cheerleader to have a unique sweater?

PURPOSE: When you have completed this lesson, you will be able to use the concepts of permutations and combinations, counting procedures, and probability to predict if something is likely to happen.

WHAT YOU NEED TO KNOW

MATHEMATICS AND PROBLEM SOLVING VOCABULARY:

1. Combinations
2. Complementary events
3. Compound event
4. Counting principle
5. Dependent events
6. Experimental probability
7. Factorial notation
8. Favorable outcome
9. Independent events
10. Mutually exclusive events
11. Permutations
12. Possible outcomes
13. Probability
14. Sample space
15. Simple event
16. Theoretical probability
17. Tree diagram

MEMORIZE THESE FACTS AND FORMULAS:

1. **Probability** tells the likelihood that something will happen, but not when it will happen.
   - The probability is 0 when an event is impossible. The closer a probability is to 0, the less likely it is to occur.
   - The probability is 1 when an event is a certainty. The closer a probability is to 1, the more certain it is to occur.
   - If the probability is $\frac{1}{2}$, it is equally likely or unlikely to occur.

2. The **probability of an event** happening = (the number of favorable outcomes) ÷ (the number of possible outcomes).

3. The number of **combinations** = (the number of permutations) ÷ (the number of different orders).

4. **Sample space** is a list of all possible outcomes of an event. Find the sample space by either making a list or by constructing a **tree diagram**.

BE ABLE TO PERFORM THESE OPERATIONS:

Finding permutations and combinations

1. **Find all possible permutations of a group of items.** A **permutation** changes the arrangement of items to form another group.
   Example #1: Find all the permutations for the word ICE.
   Solution A: If you just want to know the number of permutations, use the concept of **factorial** (3! = 3×2×1). Use 3! since there are 3 letters in ICE.
   Answer A: There are 6 permutations for the word ICE.
   Solution B: You can organize a list of permutations by keeping a letter fixed and looking at the other possibilities.
   First, start with 'I' and keep it in a fixed position. ICE, IEC
   Then, put 'C' in the first position. CIE, CEI
   Then, put 'E' in the first position. ECI, EIC
   Answer B: There are six permutations of ICE.
   ICE, IEC, CIE, CEI, ECI and EIC

2. **Use the counting principle to find all possible ways that separate events can occur together.**
   Example: How many possible ways can 5 different vehicles with 12 different colors occur?
   Answer: Take the product of the two separate events (5 vehicles × 12 colors). There could be 60 different vehicles based on style and color.

3. **Find all possible combinations of a group of items.** A **combination** is a group of events or terms. Changing the order does not change the group and does not make a new combination or group.
   Example: When making an ice cream sundae, you have the following choices to make: chocolate or vanilla ice cream, hot fudge, caramel or strawberry topping, with or without peanuts. How many different sundaes could you make?
   Solution: 2 favors of ice cream × 3 choices of topping × 2 choices of peanuts (with or without)
   Answer: 12 different sundaes

Finding probabilities

1. **Solve for the probability of a simple event.** The **probability of a simple event** happening = (the number of favorable outcomes) ÷ (the number of possible outcomes).
   Example: There are 6 pairs of socks (3 white, 1 red, 1 gray, 1 brown) in your dresser drawer. What is the probability of grabbing the red pair? a white pair?
   Answers: probability of the red pair = $1 \div 6$ or $\frac{1}{6}$, and probability of a white pair = $3 \div 6$ or $\frac{1}{2}$

2. **Solve for the probability of compound events that are mutually exclusive (two events that cannot both occur).** The probability of one or the other event happening (probability of X or Y) is the sum of the probabilities (probability X + probability Y).
   Example: A spinner has 6 equal regions numbered 1, 2, 3, 4, 5 and 6. What is the probability of spinning an odd number or a 6?
   Solution: The probability of spinning an odd number is $\frac{3}{6}$ or $\frac{1}{2}$. The probability of spinning a 6 is $\frac{1}{6}$.
   Answer: The probability of either event is $\frac{1}{2} + \frac{1}{6} = \frac{2}{3}$

Meeting the Challenge...Mathematics

3. **Solve for the probability of independent and dependent compound events (both events will occur).** It is important to see how the events are related.
    - Case 1: Two events are **independent** when the results of one event do not affect the outcome of the other. The probability of independent events C and D = probability(C) × probability(D).
      Example: A school poll of sophomores shows that 50% watch soap operas and 3 out of 4 like pizza. Find the probability that a sophomore likes soap operas(C) and pizza(D)?
      Solution: These are independent events. Probability (likes C and D) = P(likes C) × P(likes D)
      Answer: $P(C \text{ and } D) = 50\% \times \frac{3}{4} = \frac{1}{2} \times \frac{3}{4} = \frac{3}{8}$
    - Case 2: Two events are **dependent** when the outcome of one event does affect the outcome of the other. The probability of dependent events E and F = probability(E) × probability(F when E happens).
      Example: There are 6 pairs of socks (3 white, 1 red, 1 gray, 1 brown) in your dresser drawer. Find the probability of first grabbing a white pair(E) and then the gray pair(F).
      Solution: These are dependent events since grabbing the white pair changes the number left to choose from for the gray pair.
      Answer: $\text{Probability}(E \text{ and } F) = \frac{3}{6} \times \frac{1}{5} = \frac{1}{10}$

4. **Compare theoretical and experimental probabilities.** **Theoretical probability** results are derived from a formula. **Experimental probability** results come from actual experimentation. Normally, **experimental probability** gets closer to **theoretical probability** as the number of attempts in an experiment increases.
   Example: Use theoretical and experimental techniques to find the probability that a blue ball will be pulled out of the bag containing a red, a white and a blue ball. A student assistant pulled a blue ball out of the bag 38 out of 100 times.
   Solution: Theoretically, the probability would be 1 out of 3 possibilities, $\frac{1}{3}$, or approximately 33 out of 100 tries. Experimentally, the probability would be $\frac{38}{100}$ or 38 blue balls out of 100 tries.
   Answer: The experimental approach is fairly close to the theoretical.

Interpreting probabilities

1. **Determine the likelihood (probability) of an event from a fraction, decimal, percent or description.**
   Probability is a number that tells the likelihood that something will happen, but not when it will happen.
   Example: There is a $\frac{1}{3}$ probability that a blue ball will be pulled from a bag. Describe the likelihood of this event happening.
   Solution and answer: Since the probability is between 0 and $\frac{1}{2}$, it is unlikely a blue ball will be pulled from the bag.

2. **Determine if two events are complementary.** Two **mutually exclusive events** that are the only possible results are called **complementary events**. The sum of the probabilities of complementary events is 1.
   Example: A spinner has 5 equal regions numbered 1, 2, 3, 4, and 5. Find the probability(A) of spinning a 1, 2 or 3 and find the probability(B) of spinning a 4 or 5. Are these complementary events?
   Solution: Probability(A) = $\frac{3}{5}$ Probability(B) = $\frac{2}{5}$
   Answer: The sum of P(A) and P(B) = 1. These are complementary events.

BE ABLE TO PERFORM THESE CALCULATOR OPERATIONS:

Use your calculator as needed to perform basic operations.

GUIDED PRACTICE

A) Use these answers to describe the likelihood of each of these events happening.

   a) totally impossible, b) unlikely, c) equally likely and unlikely, d) likely, e) absolutely certain

   1) The weather forecast says there is a 40% likelihood for snow.

   2) There is a 100% chance of rain and it is starting to drizzle.

   3) In Ohio the days will get longer in the fall.

   4) John has a 75% chance of making the soccer team.

   Answers:
   1) ____   2) ____
   3) ____   4) ____

B) Find the number of outcomes.

   5) Cindy is taking 3 pairs of slacks (white, blue and gray) and 4 t-shirts (green, dark blue and black) with her on a trip. How many different outfits could she wear? ____

   6) Make one choice from each group below.
      Group 1: chicken, ham or fish
      Group 2: fruit or tossed salad
      Group 3: french fries or rice

      How many different outcomes are there for lunch? ____

   7) Kate can have peach, cherry or apple pie with either vanilla or butter pecan ice cream. How many possible outcomes are there? ____

C) Find the probability that each event will occur.

   8) A bag of kazoos holds 25 red, 15 blue, 20 green and 12 white. What is the probability of picking:

      a) a red kazoo? ____

      b) a blue kazoo? ____

      c) a green kazoo? ____

      d) a white kazoo? ____

   9) There are 8 freshmen, 12 sophomores, 10 juniors and 15 seniors in the pep band. One member of the pep band will be randomly chosen to represent the school at a special event. Find the probability that the student chosen is a:

      a) freshman ____   b) sophomore ____

      c) junior ____   d) senior ____

Meeting the Challenge...Mathematics

D) Vicki has 5 white, 4 black, 1 brown, 2 red and 3 blue pairs of socks in her drawer. If Vicki just grabs a pair of socks from her drawer, find the probability for each of the following situations.

10) Vicki grabs a white pair of socks. _____

11) Vicki grabs a blue pair of socks. _____

12) Vicki grabs a black pair of socks. _____

13) Vicki grabs a red pair, but one sock has a hole in it so she puts the red pair aside. Now what is the probability that Vicki grabs a brown pair? _____

E) Bob has a $20 bill, two $10 bills, one $5 bill and 7 $1 bills in his wallet. Find the probabilities.

14) He randomly picks a $20 bill. _____

15) He randomly picks a $10 bill. _____

16) He randomly picks either a $10 or $5 bill. _____

17) He randomly picks either a $1, $5 or $10 bill. _____

F) Neal's key ring has 2 house keys, 3 car keys, 1 post office box key, and 1 locker key. Find the probabilities.

18) In the dark, he needs one of the house keys. _____

19) In a hurry, he needs his locker key. _____

20) His key ring broke and scattered his keys on the floor. One of the keys fell through a grate. What was the probability the key was his post office box key? _____

INDEPENDENT PRACTICE

A) Use the calendar to find the probability of these events.

1) The date is the tenth. _____

2) The date is on a Monday. _____

3) It is after the twenty-second. _____

4) It is before the eleventh. _____

5) It is an odd-numbered date. _____

APRIL						
Sun	Mon	Tues	Wed	Thurs	Fri	Sat
	1	2	3	4	5	6
7	8	9	10	11	12	13
14	15	16	17	18	19	20
21	22	23	24	25	26	27
28	29	30				

B) Use any method to find the number of outcomes.

6) The Noontime Deli sells 12 different salads and 9 different soups during the course of a week. How many different combinations of one salad and one soup can be chosen? _____

7) Each spinner is spun once. How many different outcomes are possible? _____

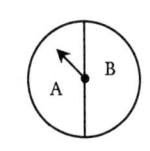

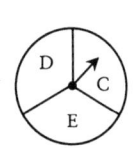

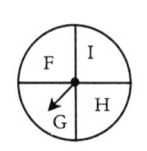

© 2003 Orange Frazer Press. All Rights Reserved.

Meeting the Challenge...Mathematics

8) Each spinner is spun once. How many different outcomes are possible? _____

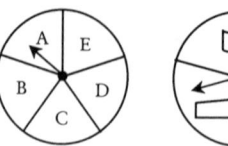

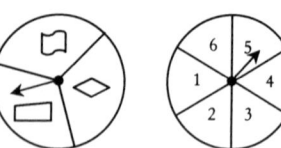

9) The Dairy Whip has 3 size cones, 5 flavors of ice cream and 6 flavored toppings. How many different ice cream cones are possible with a single topping? _____

C) Find the probabilities for the following problems.

A bag contains 5 blue balls, 2 white balls and 4 red balls.

10) Find the probability of picking a red ball out of the bag. _____

11) Find the probability of picking a white or blue ball out of the bag. _____

12) After taking a red ball out, find the probability of picking a blue ball out of the bag. _____

13) Find the probability of picking a blue ball and then a red ball. _____

14) Find the probability of picking a blue ball or a red ball or a white ball. _____

There is a 75% chance that it will rain in Dayton today and a 50% chance in Cleveland.

15) What is the probability that it will rain in Dayton or Cleveland today? _____

16) What is the probability that it will rain in both Dayton and Cleveland today? _____

17) What is the probability that it will not rain in either place? _____

TEST PRACTICE

MULTIPLE-CHOICE QUESTIONS:

1) There are 18 girls and 12 boys in a class. One class member will be chosen at random to attend a student government meeting. What will be the probability that a girl is chosen?

   a. $\frac{3}{5}$ b. $\frac{3}{2}$ c. $\frac{2}{3}$ d. $\frac{5}{3}$

2) A dish of M&Ms holds 14 brown, 8 red, 16 green, and 9 blue. Mike picked out all the red M&Ms and ate them. If Chip picks one without looking, what is the probability that Chip will get a blue one?

   a. $\frac{9}{47}$ b. $\frac{9}{14}$ c. $\frac{3}{8}$ d. $\frac{3}{13}$

Meeting the Challenge...Mathematics

3) There are four different routes from Town A to Town B. There are seven different routes from Town B to Town C. Tell how many different routes a person could travel from Town A to Town C.

    a. 11   b. 24   c. 14   d. 28

4) Jake and Marty were given the same four books from which to make a book report. What is the probability that they will report on the same book?

    a. $\frac{1}{32}$ b. $\frac{1}{16}$ c. $\frac{1}{8}$ d. $\frac{1}{4}$

5) Janelle said the probability that Alex will ask her to the prom is $\frac{1}{3}$ and the probability that Keith will ask is $\frac{1}{2}$. What is the probability that either Alex or Keith will ask Janelle to the prom?

    a. $\frac{1}{3}$ b. $\frac{2}{5}$ c. $\frac{1}{6}$ d. $\frac{5}{6}$

SHORT ANSWER QUESTIONS:

6) There are seven soccer teams in the Southeastern Soccer League. Each year every team plays every other team in the league. Find the minimum number of league games that should be scheduled. Explain your answer. Show formulas, steps and any other work to find the answer.

7) Cindy put $10 into one of the 6 pockets in her jeans and jackets, but she can't remember which pocket.

    a) What is the probability of finding the money in the first pocket she checks? _____

    b) After checking two pockets without finding the money, what is the probability of finding it in the next pocket? _____

    c) Explain what happens to the probability as she continues to check pockets for the $10.

8) A principal identifies school lockers with a code consisting of a letter followed by a numeral, such as S-3.

    a) How many codes are possible if any letter and the digits 1 through 9 may be used?

    b) How many codes are possible if the letters A – L and the digits 1-5 are used?

    c) How many codes are possible if the digits 6 – 9 are used and all letters except O, I, and Z?

    d) Explain the process used to solve parts a through c.

EXTENDED RESPONSE QUESTIONS:

9) A cooler contains 12 bottles of cola, 5 bottles of water, 8 bottles of juice and 9 bottles of iced tea. First, Peg randomly grabs a bottle of water to drink. David comes by a little later and randomly grabs a bottle of iced tea. Ten minutes later Katie randomly grabs two bottles of cola. Find the probability for each event. Describe how each event modifies the probability of the next. Discuss how the data continues to change as more bottles are removed from the cooler. Use examples, sketches, formulas and formula substitutions.

10) A cat was taught to arrange three cards in a row. One card had a 'W' written on it, another a 'T' and the third an 'O'. **Out of 104 tries, the cat actually spelled a word 28 times.** Discuss and define theoretical probability and experimental probability. Explain their differences. Find both the theoretical and experimental probabilities for this problem. Discuss the merits of both results and tell if you are satisfied with them. Discuss the experiment and make suggestions for improvement.

CLOSURE

The choices for the cheerleading sweaters include:

a) a blue or yellow sweater

b) white, black or green lettering

c) traditional or contemporary flag logo

A tree diagram was created to show all the choices.

Answer: There are exactly twelve outcomes. Every cheerleader will have her own unique sweater.

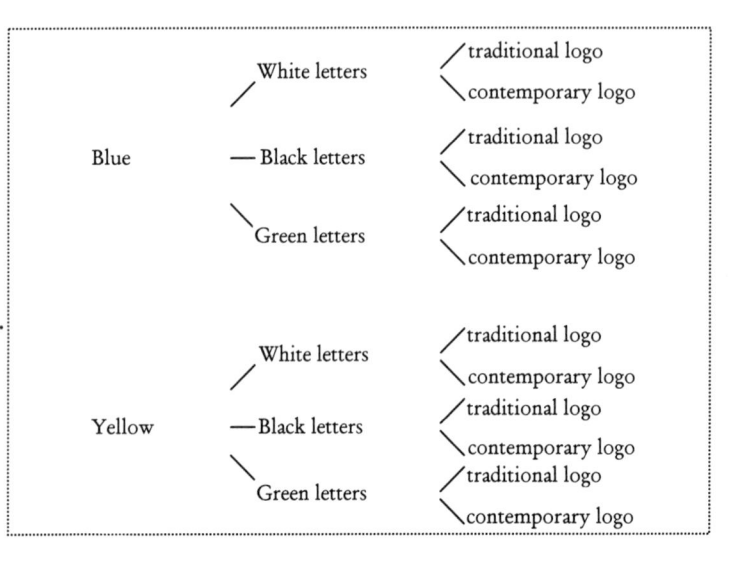

Blue
- White letters → traditional logo / contemporary logo
- Black letters → traditional logo / contemporary logo
- Green letters → traditional logo / contemporary logo

Yellow
- White letters → traditional logo / contemporary logo
- Black letters → traditional logo / contemporary logo
- Green letters → traditional logo / contemporary logo

Meeting the Challenge...Mathematics

MATHEMATICAL PROCESSES
Using Mathematics

PURPOSE: In this lesson you will apply mathematical skills to other content areas and to real life situations.

A) The Jansen Electronics Store gives a 10% discount for paying your bill within the first 10 days of the month. Bart has been saving his earnings from a part time job to buy a new game station and just saw that it went on sale at 25% off. In order to have enough money for his purchase, Bart will need to take advantage of both discounts.

1) Explain how the double discount works. _____
_____

2) Give the name and price of the most recent game station? _____, $ _____

3) What is the present sales tax rate in your county? _____ %

4) Determine the total cost of the most recent game station. Include sales tax at the current rate and both discounts.

   total cost $ _____

5) Bart has saved his last 4 paychecks of $38.53, $65.95, $43.22 and $58.18 for the purchase.

   a) Does he have enough money to make the purchase and pay for it before his next paycheck? _____

   b) If not, he will not be able to take advantage of the sale discount. Discuss some ideas of what he might do that would allow him to make his purchase and get both discounts.

B) The high school band director is looking for a new location for summer band camp. He must choose the camp with the least cost per member for his 75 band members. The price packages are listed below. Find the cost per member and put the amount in the table.

Sweet Harmony Camp	Marching Notes Camp	In Step Camp
Facilities charge: $200	Facilities charge: none	Facilities charge: $100 + $1.50 per member
$90 per member	$105 for the first 50 members, $75 for the next 50 members	$85 per member
Transportation cost: $3.50 per member	Transportation cost: $4.25 per member	Transportation cost: $2.00 per member
Cost per member: $ _____	Cost per member: $ _____	Cost per member: $ _____

The band is going to _____ Camp this summer.

© 2003 Orange Frazer Press. All Rights Reserved.

C) The Smith Mansion has a square shaped garden with an area of 625 square feet. In order to provide wheelchair access, the path around the garden must be widened and paved.

1) Find the dimensions of the garden and label the drawing.

   Dimensions _____

2) The path around the garden has uniform width (w).

   Give the dimensions (algebraically) of the entire garden and path complex. Label the drawing.

   _____

3) The area of the complete complex including garden and path is 1369 square feet. Use the quadratic formula and one other method to find the width (w) of the path.

   width of the path _____

4) Give the dimensions of the entire garden and path complex. _____

D) Jenny currently works 25 hours per week earning $6.85 per hour working at John's Burger Shack. She has been offered a job at Carole's Boutique working 25 hours per week earning $4.50 per hour plus 12% of her sales.

1) Discuss the important points that Jenny should consider before making the decision to change jobs.

2) Jenny has been assured that her sales per week will range from $500 to $750 during most of the year. During the Christmas shopping season, sales will be even higher.

   Find the range of her earnings. _____

3) Should Jenny accept this new job? _____ Justify your answer.

MATHEMATICS FORMULA AND FACT SHEET

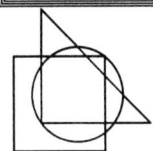

A formula sheet similar to this one will be provided to all students during the OGT Mathematics Test. Many of the following formulas will be included on that sheet, but your success on the test will depend on your knowledge of how to *use* these formulas.
Do not be totally dependent on the formula sheet. Keep the following thoughts in mind:
- Become **very** familiar with the formulas and when to use them.
- Know how to evaluate any formula.
- Be able to solve for any missing variables in a formula.

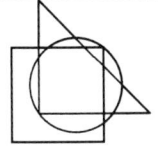

Area Formulas

parallelogram	$A = bh$		trapezoid	$A = \frac{1}{2} h (b_1 + b_2)$
rectangle	$A = lw$		triangle	$A = \frac{1}{2} bh$
circle	$A = \pi r^2$			

Circumference of a Circle
$C = \pi d$ *or* $C = 2\pi r$

Distance Formula
$d = \sqrt{(x_1 - x_2)^2 + (y_1 - y_2)^2}$

Note:
$\pi = 3.14$ or $\frac{22}{7}$

Quadratic Formula
$x = \frac{-b \pm \sqrt{b^2 - 4ac}}{2a}$

Volume and Surface Area Formulas

	Volume	**Surface Area**
cone	$V = \frac{1}{3} \pi r^2 h$	N/A
cylinder	$V = \pi r^2 h$	$SA = LA + (2 \times B)$
pyramid	$V = \frac{1}{3} Bh$	$SA = LA + B$
rectangular prism	$V = lwh$	$SA = LA + (2 \times B)$
right prism	$V = Bh$	$SA = LA + (2 \times B)$
sphere	$V = \frac{4}{3} \pi r^3$	N/A

Note:

B = area of base

LA = lateral area

Trigonometric Formulas

$\sin \angle A = \frac{opposite}{hypotenuse}$ $\cos \angle A = \frac{adjacent}{hypotenuse}$ $\tan \angle A = \frac{opposite}{adjacent}$

© 2003 Orange Frazer Press. All Rights Reserved.

MEASUREMENT STUDY SHEET

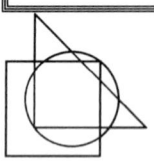

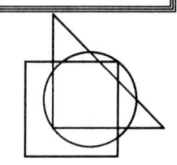

U.S. Customary Measurement

LENGTH	LIQUID	WEIGHT
1 foot = 12 inches	1 cup = 8 fluid ounces	1 pound = 16 ounces
1 yard = 3 feet	1 pint = 2 cups	1 ton = 2000 pounds
1 mile = 5280 feet	1 quart = 2 pints	
	1 gallon = 4 quarts	

Metric Measurement

LENGTH	LIQUID	WEIGHT
1 centimeter = 10 millimeters	1 centiliter = 10 milliliters	1 centigram = 10 milligrams
1 meter = 100 centimeters	1 liter = 100 centiliters	1 gram = 100 centigrams
1 kilometer = 1000 meters	1 kiloliter = 1000 liters	1 kilogram = 1000 grams

Selected Conversion Facts
U.S. Customary Units to Metric Units

LENGTH	LIQUID	WEIGHT
1 inch = 2.54 centimeters	1 quart = .95 liters	1 ounce = 28.35 grams
1 mile = 1.61 kilometers	1 gallon = 3.79 liters	1 ton = 907.18 kilograms

Temperature Conversion

Centigrade
$$°C = \frac{5}{9}(°F - 32)$$

Fahrenheit
$$°F = \frac{9}{5}°C + 32$$

GLOSSARY OF MATHEMATICAL TERMS

The number in parentheses refers to the lesson(s) where the term is found.

acute angle (6) angle whose measure is less than 90°

adjacent angles (6) angles having same vertex and common side

algebraic expression (10) numbers, symbols and variables used to express operations

algebraic term (10) an expression using numbers, variables and exponents to indicate a product or quotient

alternate exterior angles (6) given parallel lines, an outside pair of non-adjacent angles on opposite sides of a transversal

alternate interior angles (6) given parallel lines, an inside pair of non-adjacent angles on opposite sides of a transversal

altitude (5, 7) a line segment drawn from a vertex perpendicular to the opposite side

angle bisector (6) a line or segment that divides an angle into two equal adjacent angles

ascending order (1) to order from smallest to largest

average (14) a number that describes a data set (mean, median and mode)

average deviation (14) the amount that each data item differs from the mean

bar graph (13) graph comparing data using solid bars horizontally or vertically

bisector (7) a segment that divides something (segment, angle, figure, etc.) into two equal parts

box-and-whisker plot (13) a diagram or graph that shows the median and quartiles for a set of data

central angle (4, 8) an angle with its vertex at the center of a circle

chord (8) a line segment with both endpoints on the circle

circle (8) all points a fixed distance from the center

circle graph (13) compares quantities as parts of a whole

circumference (5, 8) the perimeter of a circle

combinations (15) group of items in no particular order

complementary angles (6) two angles whose sum is 90°

complementary events (15) two or more mutually exclusive events that together cover all possible outcomes

compound event (15) a combination of simple events

congruent (7) having the same size and the same shape

constant of variation (9) k, in $y = kx$

constant term (12) a variable with a fixed value

convert (1) to change one expression to another

coordinates of a point (12) an ordered pair on a graph

corresponding angles (6) given parallel lines, angle pairs that alternate on the same side of the transversal

corresponding parts (7) pairs of angles or sides in two figures that are related

cosine (4) in a right triangle, the ratio of the leg adjacent to the angle to the length of the hypotenuse

counting principle (15) a method used to determine the number of outcomes in multiple events

cross multiply (3) finding the products of means and of extremes in a proportion

degree of an equation (12) the largest exponent in an equation

degree of polynomial (10) the largest exponent in a polynomial

denominator (1) the number below the line in a fraction

dependent events (15) when the outcome of one event affects the outcome of another event

dependent variable (9) the value that is affected by the value of another variable

descending order (1) to order from largest to smallest

diagonal (5) a line segment that connects two vertices of a polygon, but is not a side

diameter (8) a chord passing through the center of a circle

direct variation (9) a function of the form $y = kx$, where k is a non zero constant

GLOSSARY OF MATHEMATICAL TERMS

discount amount (3) amount deducted from the original price to obtain the sale price

discount rate (3) a percent to be deducted from the original price

distributive property (10, 11) $a(b+c) = ab + ac$

domain (12) the possible values for x in an expression

equilateral triangle (7, 8) a triangle with three equal sides

equivalent expressions (10, 11) expressions representing the same number or value

equivalent numbers (1) different ways to express the same value

equivalent ratios (3) ratios that have the same value

evaluate (10) to find the value of an expression

exponential numbers (1) numbers written with exponents

experimental probability (15) based on a series of trials

factor (10) an integer that divides evenly into another

factorial notation (15) used in probability, $5! = 5 \times 4 \times 3 \times 2 \times 1$

favorable outcome (15) in probability, a desirable result

finance charge (2, 3) a charge connected to buying on credit

formula (11) a mathematical rule or statement

gratuity or tip (3) a percent of money given for good service

grouping symbols (parentheses, brace, bracket, fraction bar) (10) symbols used to group an expression

hexagon (8) a six-sided polygon

histogram (13) a graph where the labels for the bars are numerical intervals

horizontal or x-axis (9, 12) in a graph, the line parallel to the horizon

hypotenuse of a right triangle (7) the longest side and the side opposite the right angle

independent events (15) two events in which the outcome of the first event does not affect the outcome of the second event

independent variable (9) a variable that is not affected by the value of another variable

inequality (11) a mathematical statement that compares two expressions that are not equal

installment loan (3) money borrowed and paid back at regular intervals over a specified period

interest amount (3) a charge for borrowing money on time

interest rate (3) a percent of the amount of borrowed money

intersecting lines (6) lines that meet or cross

irrational numbers (1) numbers that cannot be written as a ratio

isosceles triangle (7, 8) a triangle with two equal sides and two equal angles

lateral area (5) the area of a lateral face

lateral edge (5, 8) the side of a lateral face

lateral face (5, 8) in prisms, the sides that connect with the base

line of reflection or symmetry (8) a line that divides a figure into mirror images

linear equation (11, 12) an equation with one solution

linear system (11) two linear equations with two variables

linear variation (9) see **direct variation**

line graph (13) a graph that shows change and the direction of change

mean (14) an average where the sum of the items is divided by the number of items

means and extremes (3) the inner two terms and the outer two terms in a proportion

measures of central tendency (14) a number that best describes a set of data

median (13, 14) the middle value in an organized set of data

median (segment) (8) the segment connecting the midpoints of the legs of a trapezoid

midpoint (6) a point on a line that divides it into two congruent segments

major arc (4, 8) an arc, greater than a semicircle, with endpoints on a central angle

GLOSSARY OF MATHEMATICAL TERMS

minor arc (4, 8) an arc, less than a semicircle, with endpoints on a central angle.

mode (14) a number that occurs most often in a set of data

monomials, coefficient of a monomial; base and exponent of a monomial (10) a single algebraic term and its components

mutually exclusive events (15) two events that cannot occur at the same time

nonlinear function (12) a function that is not a line

numerator of a fraction (1) the top number in a fraction

numerical coefficient (10) the constant value in a term

obtuse angle (6) an angle with a measure between 90° and 180°

octagon (8) an eight-sided polygon

opposite (10, 11) to change a sign in a number or expression

ordered pair (11, 12) two numbers used in a certain order to locate a point on a coordinate plane

origin (9, 12) the point where the x-axis and the y-axis intersect

outlier (14) a number in a set of data that is much larger or smaller than most of the other numbers

parallel lines (6) in a plane, two lines that do not intersect

parallelogram (7, 8) a quadrilateral with opposite sides equal and parallel

pentagon (8) a five-sided polygon

perimeter (5) the distance around the outside of a figure

permutations (15) possible orders or arrangements of a set of items or events

perpendicular lines (6) two intersecting lines forming right angles

pictograph (13) a graph that compares data using pictures or symbols

polygon (6, 7, 8) a closed figure with line segments as sides

polynomial (10) an expression that contains one or more terms

possible outcomes (15) all possible results in a probability situation

postulate (7) a statement accepted without proof

prism (8) a solid figure with two congruent and parallel faces that are polygons and other faces that are parallelograms

probability (15) the chance of an event occurring or not occurring

proportion (3, 7) two equivalent ratios

protractor (4) a tool used to measure and construct angles

Pythagorean theorem (5) $a^2 + b^2 = c^2$, where the sum of the squares of the length of the legs is equal to the square of the length of the hypotenuse

quadrants (12) the four sections on a coordinate plane separated by the x-axis and the y-axis

quadratic equation (12) a second degree equation

quadrilateral (7, 8) a four-sided polygon

quartiles (13, 14) along with the median, divides an ordered set of data into four groups of about the same size

radius (8) a segment from the center to a point on the circle

range (12, 14) the difference between the largest and smallest values in a set of numbers

rates (3) ratios that compare different kinds of units (miles per hour)

ratio (3, 7) a comparison of two numbers using division

rational numbers (1) numbers that can be expressed as ratios of two integers

real numbers (1) the set of numbers that contains all the rational and irrational numbers

reciprocal (10, 11, 12) two numbers with a product of one

reflection (8) a mirror image across a line of symmetry

regular polygon (8) a polygon with all sides congruent and all angles congruent

rhombus (7, 8) a parallelogram with four equal sides

right angle (6) an angle whose measure is 90°

right triangle (7) a triangle with one right angle

© 2003 Orange Frazer Press. All Rights Reserved.

GLOSSARY OF MATHEMATICAL TERMS

roots of an equation or function (12) solutions of the equation or function

rotation (8) a turn of a figure about a fixed point

sample space (15) a list of all possible outcomes

scale drawing (3) an enlarged or reduced drawing where the size depends on the scale and where the shape is the same as the actual object

scale, ratio or factor (7) a comparison between lengths of corresponding sides of two similar figures

scalene triangle (7, 8) a triangle with no equal sides

scatter plot (13) a graph that allows you to analyze two sets of data at the same time by plotting ordered pairs

scientific notation (1) numbers written as a product of a decimal number between 1 and 10, and a power of 10

semicircle (5, 8) a half circle (180°)

similar figures (7) figures with the same shape, but not necessarily the same size

similar or like terms (10, 11) terms that have the same variables to the same powers

simple events (15) one opportunity, one happening

sine (4) in a right triangle, the ratio of the length of the opposite leg to the length of the hypotenuse

slant height (5, 8) the perpendicular distance from the vertex of a pyramid to one edge of its base, or the shortest distance from the vertex of a cone to the edge of its base

slope of a line (12) the slant of a line in a coordinate plane

square (7, 8) a parallelogram with four congruent sides and four right angles

standard deviation (14) a way to describe how data differs from the mean

straight angle (6) an angle whose measure is 180°

stem-and-leaf plot (13) a method of organizing data using the greatest place value to group data

supplementary angles (6) two angles whose measures total 180°

surface area (5) the total area of all surfaces of a solid figure

tangent (4) in a right triangle, the ratio of the length of the opposite leg to the length of the adjacent leg

tangent line (8) a line that touches a circle at only one point

theorem (7) a mathematical statement which can be proved

theoretical probability (15) based on mathematical formulas

transformation (8) a rule of moving every point in a plane to a new location

translation (8) where a figure slides a given distance in a given direction

transversal (6) a line intersecting two parallel lines

trapezoid (7, 8) a quadrilateral with one pair of parallel sides

tree diagram (15) a diagram used to find all possible outcomes in a sample space

trend line (13) a 'best fit' line that comes close to connecting the points on a scatter plot

trigonometry (4) the branch of mathematics using similar right triangles to find measurements

variable (10) a letter that takes the place of a number

variability (14) the range, average deviation, standard deviation

vertex of an angle (6) a point where two lines, rays, or line segments meet to form an angle

vertical or y-axis (9, 12) on a graph, the line that is straight up and down or perpendicular to the horizon

vertical angles (6) opposite angles formed by intersecting lines that have equal measures

volume (5) the number of cubic units it takes to fill a figure

y-intercept of an equation (12) the y-coordinate of the point of intersection with the y-axis

zeros of an equation (12) solutions of an equation